T. J Biggs

Haematherapy

Present Status and Technique of the New Treatment

T. J Biggs

Haematherapy
Present Status and Technique of the New Treatment

ISBN/EAN: 9783337407933

Printed in Europe, USA, Canada, Australia, Japan

Cover: Foto ©berggeist007 / pixelio.de

More available books at **www.hansebooks.com**

ADVANCE SHEETS FOR REVIEW.

HÆMATHERAPY:

PRESENT STATUS AND TECHNIQUE OF THE

NEW TREATMENT.

COMPLIMENTS OF

T. J. BIGGS, M D

110 EAST 30TH ST., NEW YORK.

THE NEW MEDICAL EPOCH OF HÆMATHERAPY.

The closing decade of the Nineteenth Century has been signal-ized by probably the most potential and far-reaching development of Therapeutics in the history of Medicine ; and this is the subject of our *brochure*.

It is true that about the same period had brought to light a group of therapeutic methods of exceeding promise on special lines, the true scope and efficacy of which must yet be determined by long nd critical trial. This group of remedial methods—which may be termed the isopathic group, including its great archetype, vaccina-tion—aims to make the diseased human system at once the seat and ie beneficiary of the warfare of diseases, toxines and microbes against themselves and each other ; but it is yet far from the full development of the residual consequences, the status quo *post* bellum, for which we must watch with solicitous caution. This group is therefore not the crowning therapeutic evolution that we celebrate, whatever balance of true unalloyed beneficence it may be destined to strike in its perfected maturity ; nor is it analogous, or even harmo-nious, with the great principle of cure here asserted and vindi-cated by its unequivocal and invariable triumphs. The antithetical relation of the two principles may be tersely stated as that of casting out disease by disease or its products and reactions, on the one hand, as opposed to the expulsion of disease by the pure principle of health or vitality, on the other, through its universal depositary—living blood.

Nothing has been more indisputably settled by the modern sci-ence of biology than that health, or vitality, is as truly and practi-cally the antagonist as it is logically the antithesis, of disease, of every form of disease, and of every cause of disease. And again, it is one of the most brilliant, and by far the most practical, of all the demonstrations of modern biology, that the principle we call health or vitality is not only seated, as mankind have always blindly believed, in the Blood, but also, that perfect blood is its impregnable citadel, carrying an armament against which the assaults of disease, whether by its antecedent germs or its consequent toxines, are forever impotent.

Upon this great truth, it is clear that an infallible hæmatherapy might be founded, if only man could find means to realize the condition, *perfect blood*. Wanting this, the question yet remains : Can perfect blood be borrowed for man? On certain occasions, vital, if not absolutely perfect blood has been borrowed for the dying, by transfusion, and they have been saved. That was to supply deficient *quantity* of blood, and the means were obviously appropriate ; but it was not understood, as it is now in the light of recent science, that blood of superior *quality,* or vital energy, if thus borrowed and introduced, reinforcing the weak or morbid circulation of the patient, would have power to antagonize, master and expel disease with all its hosts, microbial or toxic, from either tissues, organs, or enervated functions. But this is a strict logical consequence of what we now know of the remedial qualities and resistant forces of the blood. Can this logical truth be made practical? and if so, in what manner, and to what extent? In a measure sufficient to make a great medical epoch, these questions are answered in favor of man by the actual applications of borrowed blood that will be clinically described in these pages.

It has now been abundantly demonstrated in practice, and only remains to be put into words and remembrance, that blood can be borrowed for us from the most vigorous animals, and supplied to our deficiency either of quantity or of vital power, by either of three modes of conveyance according to convenience or necessity, viz: by subcutaneous injection into the circulation; by direct absorption into the system through the sides of the alimentary canal, upper or lower ; or by topical application to any exposed or denuded tissue whatever, to which the native circulation of the patient fails to afford healthy sustenance and protection whether through insufficiency paucity and sickliness of organic cells, or lack of vital and vitalizing energy. That this is literally borrowing *life*, is only another and more startling way of stating the fact! a fact emphasized in a most remarkable way by the observations of Dr. Brakenridge of Edinburgh, who found that the vital cells of injected blood not only reinforced the circulation by their own quantity and quality, but also revitalized and enlarged the debilitated cells of the patient, and excited an immediate proliferation of new cells.

But a last practical question still remains : How to get the blood we desiderate, in its pure incorruptible essence and vitality, on such terms as will be practicable for the general use of mankind, and of the medical profession in particular ? The only solution of this problem so far attained—and one that seems as far satisfactory as can now be imagined—happened almost by accident to Dr. W. H. May of New York, in 1888. Baffled, like all his professional brethren, by an incurable ulcer, he tried the venturesome experiment of injecting around the sore a preparation of raw ox blood which had been up to that time known as a valuable invalid food only—"Bovinine." Astonished as if he had called up a spirit from the vasty deep, he beheld the Power he had invoked, charging from all sides upon the noxious area of disease with serried columns of vital corpuscles linked in tissue with the flesh behind, and weaving their encroaching parallels against the receding masses of the foe, until an absolute victory was won, and the ruddy flag of Life waved over the long-undisputed territory of death and corruption The discovery of this Power was afterwards seized upon by Dr. T. J. Biggs, an assistant surgeon in the New York Polyclinic, a man of ripe accomplishments and experience, and of rare professional enthusiasm, who developed new methods and applications from the germinal discovery of Dr. May, up to the marvellous therapeutic scope and efficacy now exhibited by his clinical reports, which make the marrow of the pages to which the attention of the profession is herewith invited, but which are evidently a bare earnest of vastly greater things to come.

NUTRITION AND HEALTH.

BY

DR. T. J. BIGGS.

Nutrition is the "ne plus ultra" or ultimate purpose of the various functions of the animal organism. It is the consummation of physiological action and is accomplished through the agency of all functions combined. In other words nutrition is life itself. Alimentation, digestion, absorption, secretion, respiration and circulation are all directed toward the same end—nutrition—and if any of these functions fail, nutrition becomes impaired or is discontinued, and without nutrition the body is without life.

The bond of union between these functions is the blood, the vital fluid in which all the elements of perfect nutrition are combined and endowed with the mysterious potency of supporting life. The two classes of elements necessary for the sustenance of nutrition, the nitrogenous and carbonaceous, are found in the blood in abundance, the former to supply muscular force, the latter bodily heat. Perfect nutrition demands an adequate supply of these substances, which the blood obtains from without, in the various articles of food, and, through the function of respiration, from the atmosphere. When the supply of either of these nutritive principles is not equal to the demand, nutrition becomes impaired and the functions dependent on the supply, as well as those on which it depends, become weakened accordingly.

MAL-NUTRITION.

Mal-nutrition results from any of the following causes:—1st. An inherent defectiveness of the nutritive functions; 2nd. The existence of some acute disease; 3rd. A constant lack of proper digestion and assimilation of food.

1st. An *inherent tendency* may be derived from one or both parents. The offspring thus endowed exhibits marked outward manifestations of mal-nutrition. He is apt to be small of stature and light of frame, easily overcome by fatigue and readily prone to disease. He has inherited from his parents a constitutional taint, which creates a soil well manured for the reception and growth of noxious germs and specific micro-organisms. Here is a child who is destined by nature to be a chronic invalid; one who goes through the period of childhood without missing one of the diseases which commonly attack the tender age, and all these maladies seem to assume the virulent or malignant type. Such a child, if it has not already inherited the fatal germs of tuberculosis, will be likely soon to acquire the bacilli of this dire disease from some outward source. Unless such a case as this be taken in hand early and afforded proper treatment, it would seem a dispensation from Providence that the inherited and diseased organism, lacking sufficient power of resistance against the various diseases, the seeds of which it is bound to come in contact with, should succumb early in life and thus destroy a dangerous focus of disease in one whose existence can but incite pity in others, with scant happiness for itself.

2nd. *Defective Nutrition as a Result of Acute Disease* is of course of a less serious nature, and apt to be of only temporary duration; but it is during this period of mal-nutrition in acute diseases, as a result of the exacting demands which the combustion of fever has made upon the residuary supply of nutrient material in the system, that the patient is prone to various complications and liable to die.

3d. There are various causes responsible for the existence of mal-nutrition as a result of *imperfect digestion and assimilation :* namely:

1. An insufficient or improper supply of food; 2. An excessive supply of food; 3. Habitual indulgence in alcoholic stimulants, opiates, &c.; and, — 4. Various chronic and functional diseases.

1. As already stated, the two principal classes of nutrient principles demanded by an animal organism for its sustenance are the nitrogenous and carbonaceous. When the nitrogenous is deficient and the carbonaceous abundant, the natural wear-and-tear of the tissues is, through lack of the proper material, improperly

repaired, and hence debility ensues. Of that description is the individual who to outward appearance is well nourished and his body well rounded with fat, yet lacks ability to resist fatigue. His muscles are not muscular but fatty, and the apparent muscular development is largely due to the interposition of adipose tissue. If on the other hand the carbonaceous principles are wanting, though the nitrogenous may be sufficient in quantity, those elements which are intended for combustion and the production of heat are so deficient that a great tendency to chilliness and sensitiveness to cold is experienced. In this case there is generally a deficiency of fatty tissue, and emaciation results, as the system tends to burn up the fat in the absence of carbonaceous materials supplied from without. This condition leads to pernicious results, as the fatty element in the system is a very important one. It is on this account that the emaciated subjects of scrofula and tuberculosis derive temporary benefit from cod-liver oil and oleaginous preparations until this deficiency is supplied.

2. Mal-assimilation as a result of over-indulgence in food is a not uncommon condition, and is more apt to be found among those who lead a sedentary life and eat for pleasure alone.

Dyspepsia, that common ailment of the overfed, is a functional disorder of the stomach and intestines, brought on by too rapid eating and overfeeding. It has been stated that the indolent dyspeptic, well clothed and well protected by woolen and silk, consumes more food than the hardy agricultural laborer. Such overfed individuals generally take too little exercise—and exercise is absolutely necessary for the proper assimilation of nitrogenous food and its proper adaptation to produce muscular vigor. Therefore the excess of nitrogenous material ingested by overfeeding is eliminated by the bowels and kidneys, while at the same time excess of carbonaceous material is deposited in the form of adipose tissue all over the body. Such an excessive deposit of fat, either generally or locally, is as a rule the result of overfeeding, but may be constitutional and hereditary. In either case it is not devoid of danger, as such an individual becomes more and more sedentary in his habits and less likely to resort to the safety valve and all-necessary practice of muscular exercise.

3. It is not necessary to dwell on mal-nutrition as a consequence of over-indulgence in alcoholic stimulants, opiates, &c., further than to state that it is the result of impaired digestion from constant stimulation and irritation to the whole digestive tract and from a reflex irritation of this system as a result of over-stimulation of the brain and spinal cord.

4. Lastly, mal-assimilation is to be considered as a result of various chronic diseases and functional disorders. This form of defective nutrition is probably the most common which the physician is called upon to treat. It appears in various modes or degrees according to the malady upon which it is dependent.

Diseases of the stomach, liver and intestines, lungs and kidneys, are all accompanied by impoverishment of the blood, on account of which the processes of digestion are improperly performed and nutrition necessarily becomes impaired.

The various functional disorders which produce mal-assimilation are of a reflex nature, acting through the nervous system, and are due to defective *innervation*.

Defective nutrition dependent upon some mild form of chronic disease or functional disorder, when slighted and left untreated may of itself be the cause of further disease of a structural nature, especially of the excretory organs, the kidneys and liver.

Deficient nutrition is often responsible for the contracting of various contagious maladies, and it may be stated that other things being equal, of two individuals coming in contact with some form of contagion, the one properly and the other improperly nourished, the latter is always more apt to contract the disease.

In regard to the acquiring of contagious diseases generally it may be said that there is a certain power of "resistance" which everyone possesses to a greater or less extent, which explains the reason why some during an epidemic fail to become infected while others contract the disease. Nutrition undoubtedly has more or less to do with this susceptibility or resistance to contagion in different individuals, for we all of us daily come in contact 'in the atmosphere and in our food, with various forms of noxious germs, and under certain conditions we fall victims to the various diseases of which they are the seeds, while more often they pass out of the system or are destroyed, having fallen on barren soil.

The study of bacteriology teaches us that micro-organisms exist everywhere. Everything we touch, eat and drink, and the atmosphere we breathe, may contain noxious germs, which propagate with inconceivable rapidity in the system when they find a proper soil. This we can readily demonstrate by cultures in our laboratories. Recent examinations in the dissecting room have shown the existence of tubercle bacilli in a large number of lungs, which apparently during life had produced no outward symptoms and left no evidences in the post mortem examination of pathological changes, or else left the marks of tubercular lesions, which under favorable conditions had spontaneously healed.

We all then, probably, in any large city replete with tuberculous subjects, at one time or another inhale into our lungs the germs of this dire disease. Why is it, then, that there are so many previously healthy individuals, with no family history of consumption, who acquire this malady, while on the other hand there is a still greater number who are fortunate enough to have resisted the invasion of the bacilli? There is but one positive answer to this question and that is that defective, improper or perverted nutrition must play a most important part in the etiology of all such diseases.

To recapitulate the different causes upon which mal-nutrition is dependent we make the following summary:

1. Defective nutrition in the young, an inherited condition from one or both parents. 2. Mal-nutrition the result of acute disease. 3. Defective nutrition as a result of improper assimilation and digestion, which may be brought on by an insufficient amount of food, over-indulgence in alcoholic or other stimulants, an excessive supply of food, and various chronic and functional diseases. In all of the above forms we have an impoverished condition of the blood, which is dependent upon different causes, and we have already shown in the foregoing statement that our subjects are not necessarily lean and spare, but on the contrary may be corpulent.

THE QUESTION OF TREATMENT.

The blood plays the "title role" in the causation of all forms of mal-nutrition, and as the blood is responsible for the proper performance of all those functions which go to produce perfect nutrition, if in defective nutrition the blood is lacking in one or all of these elements, it becomes incumbent upon a physician under these circumstances to supply in some manner this deficiency by such proper means as are at his disposal. The question of treatment is a most important one. There are various remedies which are used and spoken well of, all directed toward the same end, namely: that of supplying to the blood those materials in which it is deficient. But the nearer we come toward supplying these materials in the form in which they naturally exist in the blood itself, the more perfectly do we accomplish our object, and the less work we give the system in so doing, the better. What could more perfectly fill these requirements than blood itself, and what could be more rational than supplying to a system in which the blood is lacking in natural constituents, new blood from a healthy animal, which will contain in abundance all the elements necessary to healthy nutrition, and these already organized for direct incorporation in the tissues? It was often the practice in olden times for sickly individuals to visit the slaughter houses and obtain the freshly caught blood of a healthy animal, which was taken raw into the system. This procedure was followed by most beneficial results. Like treatment has been in use for poorly nourished children, whose lives were despaired of, with equally good results, and at the present day we often hear of invalids pursuing the same practice. Hence the saying "strong as an ox," which has undoubtedly derived its origin from this source. In the light of past experience and as a result of many practical demonstrations which have been presented, I wish to lay special stress upon the treatment of mal-nutrition by the use of bullock's blood, and I especially desire to emphasize the great value of its employment in all forms of cases as classified in the foregoing pages. Fortunately it is not necessary at the present day to send our patients to the slaughter house, as we can obtain our ox's blood fresh from the arteries of healthy animals, preserved in all its vitality in the preparation known as Bovinine.

GENERAL OUTLINES OF TREATMENT.

Defective nutrition in childhood may not be recognized in the nursery, and being consequently neglected, such children, though apparently healthy, may sicken and die of phthisis before coming to maturity. If close observation is made, defective nutrition may be discovered before it becomes too late, and by proper treatment the child can be restored to a condition of health and acquire a necessary power of resistance against the various diseases which threaten the life of so many.

In many young children the use of blood is especially to be recommended and the preparation known as Bovinine is a most eligible form in which to administer it and is, in fact, the only one that is readily to be obtained. It is perfectly reliable and stable, a pure defibrinated ox's blood free from all insoluble material. In children the proper quantity to give is 10 to 20 drops at each feeding, in milk or water, with salt added to suit the taste.

It is hardly necessary to say any more in regard to building up the young by this natural method. The intelligent attendant must see the wisdom of this mode of treatment and will follow it up according to his own ideas.

The next class of cases we have discussed are those dependent upon some acute disease. Here in the so-called convalescing stage, when the patient is worn out and weak, some powerful means of rebuilding the system is demanded. Here the value of the blood treatment will be readily appreciated and especially recommend itself to those who object to the employment of alcoholic stimulants, while it may be also used in combination with them. In the adult I would recommend from a tablespoonful to a wineglass of the Bovinine to be given in milk or light wine every two, four or six hours. It is astonishing to see the marked improvement resulting from such a course of treatment, the response following so rapidly and surely.

Under the phase of mal-nutrition last described, viz.; that resulting from constant lack of proper digestion and assimilation, there is no end to the variety of ailments which are amenable to treatment by the means here suggested. Where the supply of food has been insufficient or improper, the introduction of aliment already perfected and vitalized instantly supplies the wants of the system without exacting of it any functional effort. It is a direct supply to the system of all that constitutes healthy blood, both in its vital energy and in the nutritive elements for want of which the broken-down organism is languishing. This course of treatment, if there is any vitality left in the patient, will probably enable him to shake off his disease and regain sufficient vigor for a long and useful life.

Under this class also will come the habitual drinker or opium eater, who has ruined his digestion and powers of assimilation by such vicious indulgence. Abstinence from the alcoholic or other habit will give the blood treatment a chance to show its power in healing and reinvigorating his outraged system and enable him to experience the pleasure of good health.

In the case of those who suffer from over-eating it is especially necessary that in conjunction with treatment a proper and regular amount of physical exercise be taken. The diet must also be regulated and may be cut down to nil, as in its place we shall supply in small bulk all that is necessary to the system, concentrated in the ox blood. I can personally state that physical exercise, properly pursued, in conjunction with the intelligent use of Bovinine, may produce a condition of physical health and well-being that is only exampled in those athletes with whom physical development is the principal pursuit of life. In confirmation of this I may quote the experience of several who have followed this mode of training, and especially the remarkable case of one who in a recent bicycle race used no other mode of nourishment except the daily ingestion of Bovinine.

Many athletes in training find it beneficial to confine themselves almost entirely to Bovinine as the means of nourishment. Of course its use in such cases must be regulated in quantity according to individual physique and the effect experienced. Furthermore, some systems require a greater amount of one or another element of food, and such elements must be supplied as the demand for them becomes apparent. It will, of course, be understood, that under conditions of normal health, something of the substantial filling material provided in natural food is not to be omitted.

In chronic diseases the indications are well marked. These are always the condition and effects of mal-nutrition fostered by the disease, while the process of the disease is in turn accelerated by the want of nutrition. Such a means as the life-giving animal blood affords for rapidly building up a disabled system forcibly appeals to us in such cases and it is only necessary to state that the preparation Bovinine is to be used in the same manner as already prescribed. Even if it fails to exert a direct curative action on the malady itself, it will at least prolong life by supplying those vital elements which the various functions of the organs in their diseased condition are making unsatisfactory efforts to produce for themselves.

In conclusion I might mention a few of the special indications for the adoption of the blood treatment:—

1. That condition of extreme collapse and shock, the result of severe hæmorrhage. Here we are called upon to supply at once something to replace the great loss of a large quantity of the vital fluid with all its nutrient principles. Hence we proceed at once to inject subcutaneously into the soft parts a half a pint or more of the sterilized bullock's blood or Bovinine, (with the proper administration of one-third salt water), and the effect we produce is instantaneous. This process may be repeated, or followed up by the internal administration of the Bovinine. The Bovinine may also be administered per rectum in the above condition with most satisfactory results. Its use in this manner in that alarming condition—postpartum hæmorrhage—will be spoken of in a subsequent article.

2. In extremely irritable conditions of the stomach we have another special indication for the use of Bovinine, which can be given internally oftentimes when nothing else will be retained. Or in those cases where even the Bovinine is ejected and rectal feeding has to be resorted to, enemata composed of Bovinine and peptonized milk will contain all that is necessary to sustain the system until stomach feeding can be resumed. In such a case two or three ounces each of Bovinine and peptonized milk should be given per rectum every three or four hours, and should be introduced through a tube which will reach at least as high as the sigmoid flexure of the colon—eight inches.

This mode of treatment is adaptable to all cases where it is desirable to give the stomach a rest, where it is the seat of some disease or growth or is unduly irritable, as after the various febrile diseases and especially typhoid fever, when the local application of Bovinine to the ulcerated surface of the intestines finds a special efficacy.

3. In anæmia and chlorosis in young girls, especially those who suffer from menorrhagia, Bovinine will render invaluable aid.

4. Lastly, as a local application to all forms of ulceration, unhealthy sloughing surfaces, and as a means of preserving locally perfect nutrition during a process of skin grafting, Bovinine has proved its value by a most practical showing of cases which will be reported from carefully taken notes, in subsequent pages.

The use of Bovinine as a local application is based on the same grounds as its internal administration, and upon the fact that by this method the pure animal blood is brought directly in contact with the tissues, which is the ultimate purpose of the circulatory function.

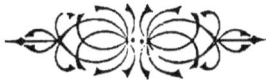

LOCAL NUTRITION IN CHRONIC ULCERATION.

Is the Topical Application of Nutrition Available in the cure of Chronic Ulceration?

By T. J. BIGGS, M. D.

ASSISTANT SURGEON NEW YORK POLYCLINIC.

The question of curing chronic ulceration has been one of the most profound and one we might say that has puzzled the surgeons of the world for centuries past, who in spite of all the skill and with all the remedies at their disposal, have in many cases failed to procure complete repair. The question then is: What is the cause of ulceration, and what is the pathological factor in its production?

Wyeth's Text Book on Surgery describes an ulcer as "the result of the molecular death in the integument or mucous membrane and the underlying areolar or sub-mucous tissue." The process of necrobiosis may at times extend below the deep fascia. Of whatever variety, an ulcer is caused by a failure of nutrition in the part affected. The arrest of nutrition may be from a specific cause, as in the ulcer of chancroid or from a varicose condition of the veins, or constitutional, as in the late manifestations of syphilis, in scorbutus, etc. Where there is dyscrasia, ulcers are more liable to occur, notably in parts of the body subjected to undue irritation, or where an abnormal interference with the circulation exists. Ulcers may be divided into two clinical groups —the *active* and *indolent*. In the one, the material for repair is in excess ; in the other it is deficient. The most frequent seat of ulcer is upon the anterior aspect of the tibia at its middle or lower portions. The indolent form occurs almost always in the aged, and chiefly among the poorly fed and laboring classes, where the erect posture is of necessity maintained for many successive hours. *An indolent ulcer demands stimulation.*

From the foregoing statement, the treatment of indolent ulceration must of necessity be directed to the cause of tissue destruction, and the application of such available nutrition as shall prevent further destruction and decomposition, to assist nature in repairing her loss.

With the view of ascertaining precisely how far the topical use of nutrition was applicable, a series of experiments were conducted and various forms of nutrition were applied to ulcerated surfaces. Dilute solutions of sugar, starch, predigested starch, malt extracts, peptone broth, beef tea, extracts of beef, dissolved Pepsin, Papoid, Ichthyol, milk, blood clot, ordinary defibrinated blood, and defibrinated ox-blood available in the preparation known as Bovinine. Ten consecutive cases were tried with each substance and the effect observed, care being taken to note any existing diathesis in each particular patient, which, if discovered was promptly treated. In most of the experiments three or more substances were used on the same patient at the same time, according to the number of the ulcers, and while all rendered more or less benefit, milk, blood clot, defibrinated blood or Bovinine were more prompt in their action than the rest. Further experiments proved that milk turned rapidly sour and produced irritation, blood clot and ordinary defibrinated blood acted very nicely when the dressings were changed every four hours, but when left longer were found to be decomposed, and thereby retarded the process of healing, which was slow compared with that of the Bovinine cases, the ratio being about three to one in favor of the latter. A series of experiments were also conducted with Bovinine and Iodoform on the same patient; here the ratio was found to be about ten to one in favor of the former.

The explanation of this process will be evident to all who are familiar with the cell theory of Virchow, or the study of embryology, which shows conclusively that all living bodies are composed of one or more cells, each cell being peculiarly adapted to perform the function ascribed to it. Among the *lower orders* of cell life, the Amœba probably are the most common. These closely resemble the Leucocytes or white corpuscles of vertebrate blood, and are almost entirely composed of undifferentiated protoplasm. On close examination, they will be found to consist of small, oval-shaped masses, which by means of a continually occurring flux in their protoplasmic substance, are able to change their form and also their position ; each one is capable of living, moving, absorbing, growing and finally dying.

Furthermore, each one of these little bodies will be found to be contractile, just as the muscle-fibres are under the action of an electric current, with this exception, that, while the muscular contraction is regular, the cell contraction is spontaneous. When irritated they will contract and produce heat which is relative to inflammation. They are receptive and assimilative, fragments of foreign matter are laid hold of by a well-defined movement, which is quite characteristic and often repeated, and the observer can often see dead spores and pieces of undigested matter imbedded in their interior, which are subsequently ejected from their bodies. They are metabolic and secretory, for the increase of the Amœba is incessant, and the protoplasm is continually undergoing chemical changes, room being made for new protoplasm by the breaking down of the old into products which are ejected. They are also respiratory and reproductive, oxygen being absorbed and the production (complete or partial) of carbonic-acid gas ensues. The original cells increase their number by cell division.

These individual cells by *aggregation into colonies,* reach their maturity and subsequent development in the matrix or mother element, in the higher order of animals and plants. Each colony or tissue corresponds to one or more of the fundamental qualities of protoplasm, for the development of which it is devoted by the special division of labor. Chemical change—including the assumption of oxygen, and the production, complete or partial, of carbonic gas, and above all, that of assimilation, secretion, excretion and reproduction— is retained by every tissue to a greater or less extent, and while some of its processes are held in abeyance for mechanical reasons, yet, it is never totally arrested while life exists. Plants acquire their nutriment from the inorganic kingdom and from decomposing organic matter ; animals, chiefly from organic matters, whether animal or vegetable. Both possess the wonderful power of re-arranging the constituents of these substances into forms identical with those of the elements of their various tissues, and of thus making them part and parcel of themselves.

Experiments have been made with these individual cells and isolated colonies of simple structures by supplying nutrition artificially, by which they have been kept alive for days independent of the influence of the nervous system, each cell assimilating by contact with the proper nutritive medium, the material necessary for the preservation of its growth and integrity. The food or aliment which is necessary for the sustenance of these cells and colonies must be in a state of solution and contain the elements of Carbon, Hydrogen, Oxygen, Nitrogen and the various salts. It is therefore proven conclusively that each and every tissue is capable of absorbing and assimilating nutrition when presented in proper form.

Upon these conditions is based the theory of topical feeding in chronic ulceration, and to this inherent power of life in each individual cell, do we owe the success attained by the topical application of nutrition as alluded to above.

Granulating surfaces are non-absorbent if the substance presented be of an acrid, irritable nature or be incapable of sustaining the life of these individual cells, but, if on the other hand, the substance presented be of a neutral reaction, of a greater density than that of healthy blood, and be composed of substances capable of sustaining the life of those cells, not only will they absorb and assimilate such nourishment in large quantities, but they will live and thrive, and assist nature in repairing her loss, in such a way as to leave the surface of the part free from scar.

The subjoined cases afford practical demonstration of the principles stated in the above disquisition.

These cases are from among the first that were treated by this method. They have been followed so far (Oct. 10, 1894), by 348 other cases of like nature, without a single failure of perfect cure.

CASE No. 1.

NAME.—Mrs. Crummey. NATIVITY.—Ireland. AGE—Fifty-five years. RELATION.—Married. WORK.—Housework. ADDRESS. —No. 303 East 35th St., N. Y. City.

Patient came to Demilt Clinic, Oct. 9th, 1893. Examination revealed a Traumatic Ulcer of right leg two inches in diameter over the gastrocnemius muscle. She gave a history of falling down stairs, sustaining a severe contusion of right calf, which broke down in ulceration two weeks later, and in spite of all treatment applied, refused to heal. The ulcer was foul and covered with unhealthy granulations, exuding a thick viscid sanguineous pus; the edges were indurated and undermined; there was also considerable phlebitis. Its failure to heal and respond to ordinary means of treatment seemed to be on account of a local malnutrition. On these grounds it was decided to resort to a means of supplying the lack of nutrition by the local application of Bovinine. The method of treatment pursued was as follows : The leg was thoroughly scrubbed with hot water and bichloride of mercury soap, the unhealthy granulations removed with a curette, and the wound was then dressed with wet bichloride gauze 1-3000, and patient instructed to call at my office next day. Following this the daily use of Bovinine was adopted exclusively, it being applied topically as follows : After thoroughly cleansing the wound with peroxide of hydrogen, plain sterilized absorbent gauze was saturated with Bovinine and laid over entire surface of the ulcer, over this a piece of sheet lint large enough to overlap the edges of the gauze at least three inches in all directions, and over this a piece of oiled muslin (it is wise to use the oiled silk or muslin in all Bovinine dressings, as it promotes a more ready absorption and prevents evaporation of the volatile elements in the Bovinine) and a roller bandage firmly applied. The ulcer was dressed daily for ten days, then every other day until discharged. The ulcer healed in just three weeks from the first application of Bovinine, leaving only a slight pink cicatrix.

T. J. BIGGS, M. D.,
Asst. Department Surgery, Demilt Clinic, New York City.
Dec. 10th, 1893.

CASE No. 2.

NAME.—Matilda Dukner. NATIVITY.—Germany. AGE.- - Twenty-six years. RELATION.—Married. WORK.—Housework. ADDRESS.—No. 715 Seventh Ave., N. Y. City.

Dr. L.—— sent patient to me October 25th, 1893. On examination I found a superficial necrosis of entire first phalanx of index finger of left hand, resulting from a periostitis, the consequence of a blow received four weeks previous. The wound was enlarged by an incision and necrosed tissue removed. After thorough irrigation the cavity was packed with sterilized absorbent gauze saturated with Bovinine, and dressed as in the previous case. The packing was renewed each day until the cavity had entirely filled up. The

wound was then strapped, and in twenty-one days from the first application of Bovinine the patient was discharged with the wound entirely healed.   T. J. BIGGS, M. D.,
Demilt Clinic.

CASE No. 3.

NAME.—John O'Brien. NATIVITY.—Ireland. AGE.—Fifty-six years. RELATION.—Married. WORK.—Flagman, L. I. R. R. ADDRESS.—No. 195 Mulberry St., N. Y. City.

Dr. F.—— sent patient to me October 3rd, 1893, suffering from a necrosis of right tibia, the result of a railroad injury. Examination revealed deep necrosis on anterior surface of tibia, about three inches long, in the middle of the shaft. The wound was rendered antiseptic, and the necrosed tissue removed, leaving a mere shell of bone. It was decided on account of the size of the cavity to try sponge grafting, which was done as follows : A thin layer, cut from a laparatomy sponge (previously prepared by soaking for three hours in a 1-40 sol. of carbolic acid, and then sterilized) was soaked in Bovinine and laid in bottom of cavity, and the wound was then covered with wet Thiersch gauze. The wound was dressed in this manner daily for five days, at the end of which time the layer of sponge had become vitalized, granulations anastomosing all through its pores. A second layer of sponge soaked in Bovinine was again applied over this granulation mass and dressed as before: in seven days this also became vitalized, filling up almost the entire cavity. From this time on until the bone had entirely covered, the wound was dressed with nothing but Bovinine. After the bone had entirely covered, the wound was strapped and dressed with dry dressings, iodoform gauze, etc. December 10th, 1893, the patient was discharged, the wound having entirely healed.
T. J. BIGGS, M. D.,
Demilt Clinic.

CASE No. 4.

NAME.—Anna Battey, NATIVITY.—England. AGE.—50 Years. RELATION.—Married. WORK.—Housework. ADDRESS.— 178 Freeman Street, Greenpoint, L. I.

Patient came to Demilt Clinic, Oct. 9, 1893. An examination revealed a varicose ulcer of left leg over ankle, on the outer side, 2½ inches in diameter and of eleven years standing.

The patient said that in spite of numerous treatments applied the ulcer had never entirely healed: it would improve for a while and then break down again.

After thoroughly cleansing the ulcer and surrounding tissues, the ulcer was curetted and a wet Thiersch dressing ordered for the first twenty-four hours, after which Bovinine topically applied as in the former cases, was the only treatment employed. The ulcer was dressed daily for the first ten days, then every other day until the patient was discharged cured.

The ulcer healed in just twenty-eight days, leaving a pink cicatrix one-half inch in diameter.

.   T. J. BIGGS. M. D.,
Asst. Department Surgery,
Demilt Clinic,
New York City.

CASE No. 5.

NAME.—Mary Byrne.   NATIVITY.—Ireland.   AGE.—40 Years.   RELATION.—Married.   WORK.—Housework.   ADDRESS. — 205 East 40th Street, New York City.

Patient was first seen at Demilt Clinic, Oct. 10th, 1893. On examination found a large varicose ulcer on left leg, above ankle on outer side, six inches long and two and one-half wide, also a smaller varicose ulcer of right leg, two inches in diameter.

The large ulcer on left leg was of fourteen years standing, the small one of three months; neither of them from the time of their first appearance had ever healed although patient had been constantly treated at various clinics.

The topical application of Bovinine was commenced and applied daily for the first two weeks, after that every other day. The small ulcer healed in twenty-five days and the large one in eight weeks, leaving in both instances pink, healthy scars. The patient has been seen twice a week since the ulcers healed, and up to date there is no indication of any return, but on the contrary the scars have grown less and the leg has assumed its normal size.

T. J. BIGGS, M. D.
Asst. Department Surgery,
Demilt Clinic,
New York City.

CASE No. 6.

NAME.—Sam. Johnston.   NATIVITY.—Ireland.   AGE.—35 Years.   RELATION.—Married.   WORK.- Machinist.   ADDRESS.— 223 Avenue C, N. Y.

DIAGNOSIS.—Indolent Traumatic ulcer of left leg, two inches in diameter, over outer side of ankle joint; ulcer of two years standing and had never been healed.

TREATMENT.—Topical applications of Bovinine commenced Oct. 9th, 1893, patient was discharged completely cured November 4th, 1893.

CASE No. 7.

NAME.—Mary Ann Callery. NATIVITY.—Ireland. AGE.—60 Years. RELATION.—Married. WORK.—Housework. ADDRESS.— 110 Sanford Street, Brooklyn, N. Y.

DIAGNOSIS.—Varicose ulcer of right leg, over ankle on outer side, 1½ inches in diameter, of three years standing: had never healed in spite of various treatments.

TREATMENT.—Bovinine commenced topically Oct. 15th, 1893, patient was discharged cured Nov. 10th, 1893.

CASE No. 8.

NAME.—Maggie Hanabery. NATIVITY.—Ireland. AGE.— 39 Years. RELATION.—Married. WORK.—Housework. ADDRESS. —514 East 17th Street, N. Y.

DIAGNOSIS.—Large indolent ulcer 4½ inches in diameter, on right foot, dorsal surface, of 17 years standing, the result of burn.

TREATMENT.—Topical application of Bovinine was commenced Nov. 10th, 1893, patient was discharged completely cured Jan. 15th, 1894. In this case the patient suffered severely and constantly until Bovinine was applied, after which she suffered absolutely no pain.

CASE No. 9.

NAME.—Lizzie Ziskind. NATIVITY.—United States. AGE.— 2 Years. ADDRESS.—411 Seventh Avenue, N. Y.

DIAGNOSIS.—Indolent ulcer 4 inches in diameter, of three months standing, on left nates, the result of a burn.

TREATMENT.—Bovinine commenced topically Dec. 16th, 1893, discharged completely cured Jan. 25th, 1894. The patient was entirely relieved of pain after first application of Bovinine.

CASE No. 10.

NAME.—Pat. Murray. NATIVITY.—Ireland. AGE.—55 Years. RELATION.—Married. WORK.—Longshoreman. ADDRESS. — 25 Rutgers Street, New York.

DIAGNOSIS.—Ulcer on left side of the lower jaw with necrosis of bone, the result of a carious tooth; ulcer 2 inches in diameter and of two years standing.

TREATMENT.—December 6th, 1893, necrosed bone was removed and wound packed with Bovinine; this was repeated every day for 2 weeks, then every other day until discharged cured February 22d, 1894. This wound in spite of three operations and all treatment employed, had refused to heal until Bovinine was used.

CASE No. 11.

NAME.—Nellie Ravora.   NATIVITY.—Ireland.   AGE.—25 Years.   RELATION.—Married.   WORK.—Housework.   ADDRESS. —208 East 25th Street, New York.

DIAGNOSIS.—Indolent ulcer of right leg, above ankle joint on the outer side; ulcer 1¾ inches in diameter.

TREATMENT.—Bovinine applications were commenced January 1st, 1894, and was discharged cured February 3rd, 1894. In this case the patient suffered severely from pains in and around the ulcerated surface, but was entirely relieved after the first application of Bovinine.

CASE No. 12.

NAME.—Harry Rimanoczy.   NATIVITY.—England.   AGE.— 25 Years.   RELATION.—Single.   WORK.—Artist.   ADDRESS.— 831 3rd Avenue, New York.

DIAGNOSIS.—Ulcer on corona of penis ½ inch in diameter and of two years standing.

TREATMENT.—Applications of Bovinine commenced October 20th, 1893, patient discharged cured November 28th, 1893. This little ulcer had resisted all treatments till Bovinine was used.

CASE No. 13.

NAME.—Kate Keller.   NATIVITY.—Ireland.   AGE.—55 Years. RELATION.—Married.   WORK.—Housework.   ADDRESS.—605 East 16th Street, New York.

DIAGNOSIS.—Varicose ulcer, outer side of knee, on right leg, ¾ inch in diameter, of six months standing.

TREATMENT.—Bovinine commenced topically October 31st, 1893, was discharged cured December 1st, 1893.

CASE No. 14.

NAME.—Bridget Welsh.   NATIVITY.—Ireland.   AGE.—65 Years.   RELATION.—Married.   WORK.—Housework.   ADDRESS. —295 Delancy Street, New York.

DIAGNOSIS.—Varicose ulcer on right leg, outer side, just above the ankle, 2 x 3 inches, of twenty years standing; had never been entirely healed.

TREATMENT.—Commenced Bovinine November 8th, 1893, patient discharged December 25th, 1893, cured. In this case as in all the others the patient was entirely relieved of pain after the first application of Bovinine.

CASE No. 15.

NAME.—Mat. McGray. NATIVITY.—Ireland. AGE.—38 Years. RELATION.—Married. WORK.—Driver. ADDRESS.— Yonkers, N. Y.

DIAGNOSIS.—Traumatic ulcer of right leg, outer side, just above ankle, 2 inches in diameter, of sixteen years standing and had never healed.

TREATMENT.—Bovinine dressings begun November 9th, 1893, patient discharged cured December 16th, 1893.

CASE No. 16.

NAME.—Margaret Donovan. NATIVITY.—Ireland. AGE.— 50 Years. RELATION.—Married. WORK.—Housework. ADDRESS. 7 Oak Street, New York.

DIAGNOSIS.—Varicose ulcer of right leg, 2 inches in diameter, of two years standing, had never healed.

TREATMENT—Applications of Bovinine were commenced November 13th, 1893, patient discharged cured December 19th, 1893.

CASE No, 17.

NAME.—Hannah Kirwin. NATIVITY.—Ireland. AGE.—65 Years. RELATION.—Married. WORK.—Housework. ADDRESS. —1499 Lexington Avenue, New York.

DIAGNOSIS,—Indolent ulcer on right leg, of two years standing, 2 x 1½ inches; had never been healed.

TREATMENT.—Bovinine applications were commenced December 7th, 1893, patient discharged cured January 28th 1894.

CASE No. 18.

NAME.—Louise Gobeley. NATIVITY.—Germany. AGE.—24 Years. RELATION.—Married. WORK.—Housework. ADDRESS. —27 Broome Street, Newark, N. J.

DIAGNOSIS.—Tubercular ulcer three by two and one-half inches in diameter, on right leg over tibia, of six years standing, had never healed in spite of various treatments applied.

TREATMENT.—March 15th, the ulcer was curetted, and then injected subcutaneously with the mixture of Bovinine and Iodoform at the line of demarcation between healthy and unhealthy tissue, thirty minims of the mixture were injected in six different places, the wound was then cleansed and dressed with Bovinine as in the other cases. The dressings were renewed daily for the first eight days, then every other day, until the patient was discharged entirely well April 10th, 1894.

24

CASE No. 19.

NAME.—Emma Robbins. NATIVITY.—United States. AGE.
30 Years. RELATION.—Single. WORK.—Dressmaker. ADDRESS.
—441 W. 47th Street.

DIAGNOSIS.—Patient came to Polyclinic December 2nd, 1893.
An examination revealed a necrosis of the first phalanx of the
index finger, the result of an injury. The necrosis involved almost
the entire anterior surface of the phalanx, the bone being completely denuded of its periosteum.

TREATMENT.—After the necrosis had been removed, the wound
was packed with Bovinine; the wound was re-dressed daily for the
first seven days, then every other day until healed, Jan. 6th, 1894.

CASE No. 20.

NAME.—Henry Coligan. NATIVITY.—Ireland. AGE.--76 Years.
RELATION.—Married. WORK.—Machinist. ADDRESS.—324½ East
8th Street.

DIAGNOSIS.—Traumatic indolent ulcer of right leg, over outer
side of ankle joint, about the size of a twenty-five cent piece. In
spite of all treatments applied it would not heal; the ulcer was of
six years standing.

TREATMENT.—Hypodermic injection of Bovinine and salt water
(one-fourth salt water, three-fourths Bovinine) was commenced
May 10th, 1894, and employed twice a week for the first fourteen
days, then discontinued, the ulcer having almost healed; besides the
injections the ulcer was dressed topically with Bovinine, as in the
other cases described. The patient was discharged cured June
3rd, 1894.

CASE No. 21.

NAME.—Eddie Garrison. NATIVITY.—United States. AGE.—
3 Years. ADDRESS.—217 East 47th Street.

DIAGNOSIS.—Traumatic ulcer on back of right hand, about
one and three-quarter inches in diameter, of four months standing,
had resisted all treatment employed.

TREATMENT.—Bovinine commenced topically June 3rd, 1894,
patient discharged cured July 7th, 1894.

CASE No. 22.

NAME.—Edward Miller. NATIVITY.—United States (Negro).
AGE.—22 Years. RELATION.—Single. WORK.—Hall-boy. AD-
DRESS.—217 West 27th Street.

DIAGNOSIS.—Syphilitic ulcer of right leg, two and one-half
by three inches, of two years standing, resisted all treatment.

TREATMENT.—The ulcer was curetted, and dressed with a wet bi-chloride ot mercury dressing for the first twenty-four hours, after which nothing but Bovinine dressings were employed; the first dressing was applied May 4th, 1894, and the patient discharged cured June 12th, 1894.

CASE No. 23.

NAME.—James O'Connor. NATIVITY.—Ireland. AGE.—55 Years. RELATION.—Married. WORK.—Cigar Maker. ADDRESS. —195 Third Avenue.

DIAGNOSIS.—Indolent ulcer of right leg, two by three inches in diameter, of five years standing, had never healed.

TREATMENT.—Topical applications of Bovinine were commenced June 4th, 1894, patient discharged cured July 16th, 1894.

CASE No. 24.

NAME.—Chas. McGinnis. NATIVITY.—United States. AGE. 15 Years. WORK.—School. ADDRESS.—306 East 49th Street.

DIAGNOSIS.—Traumatic ulcer, right side of head over temporal bone, one by one and one-half inches in diameter, of six months standing.

TREATMENT.—Bovinine dressings commenced May 4th, 1894, discharged cured May 29th, 1894.

CASE No. 25.

NAME.—Dan Burk. NATIVITY.—Ireland. AGE.—35 Years. RELATION.—Single. WORK.—Laborer. ADDRESS.—Washington House, 23rd Street.

DIAGNOSIS.—An ulcer the size of a twenty-five cent piece over joint of great toe, on right foot, the result of a burn. The ulcer was very painful and of eight months standing.

TREATMENT.—Bovinine commenced topically April 6th, patient discharged cured May 2nd, 1894; after the first application of Bovinine the patient was relieved of all pain.

CASE No. 26.

NAME.—Arthur Thomas. NATIVITY.—France. AGE.—38 Years. RELATION.—Single. WORK.—Cook. ADDRESS.—70 Bowery.

DIAGNOSIS.—Ulcer of hand, two and one-half by three and three-quarter inches, the result of a burn, patient suffered constantly.

TREATMENT.—Bovinine dressings commenced Jan. 12th, 1894, patient discharged cured Jan. 30th, 1894. Patient was relieved of all pain after first dressing of Bovinine.

CASE No. 27.

NAME.—Mary Breitfuld. NATIVITY.—Germany. AGE.—48 Years. RELATION.—Married. WORK.—Housework. ADDRESS. —644 Third Avenue.

DIAGNOSIS.—Varicose ulcer of left leg, over tibia, three and one-half by two and one-half inches, of six years standing.

TREATMENT.—Bovinine dressings commenced Feb. 20th, 1894, patient discharged cured April 2nd, 1894.

REMARKS.—Patient suffered so much in this case, that she wanted the leg amputated. After second dressing of Bovinine she was entirely relieved of pain.

SALEM, MASS., JUNE 1st, 1894.

THE BOVININE COMPANY,
Gentlemen:—

In regard to the *external use* of Bovinine by me, and as a local nutrient, I may say that it proved very satisfactory.

In some cases of multiple chronic ulcers of the leg, the rapidity of healing under its use was really remarkable.

Compresses soaked with Bovinine were applied, and it was also used subcutaneously. Its use was continued in various forms of ulcers, old and recent; wounds of all kinds; compound fractures; sinuses; after operations for fistula, etc., etc. In one case only some irritation was caused by its use subcutaneously; in all others it was *satisfactory and successful*

Yours truly,
LAWRENCE G. KEMBLE, M. D.

CAMDEN, N. J., January 4th, 1894.

BOVININE COMPANY, New York.
Gentlemen :

I consider it my duty to again inform you of the great value of Bovinine. I am highly gratified over one case in particular. This case of which I am about to speak is one of a lady that suffered a traumatic injury low down on the calf of her leg some three months ago, which became a painful unyielding ulcer. The lady is about 55 years of age, ordinarily very healthy ; she is a nurse by occupation. After she had been treated at the university for six weeks the ulcer became worse than when the treatment began. She came to my office about three weeks ago, I curetted the ulcer, put the limb at rest and applied Bovinine, which acted like magic, and by the time you have received this letter the case will be entirely well from the simple application of Bovinine. I used in treating this ulcer only about two ounces.

Please accept my thanks again and again for Bovinine.

Yours truly,
Dr. W. G. BAILEY.

Through the courtesy of Dr. B—— of Chicago, we are allowed to cite the following two cases, illustrations of which will be found on pages 28 and 29.

These cases were treated with Bovinine by Dr. B—— with remarkable success.

CASE No. 1.

NAME.—John J. AGE.—50 Years. RELATION.—Married. WORK.—Mechanic. NATIVITY.—United States.

The patient was first seen by Dr. B—— August 1st, 1893. An examination revealed an extensive ulceration on the left fore-arm, the result of a burn received five years previous, which the patient said had never healed in spite of all treatments applied. When first seen the ulcer was covered with a thick, viscid, sanguineous discharge, of very foul odor, underlying which was an unhealthy granulating surface.

TREATMENT.—The ulcer was thoroughly cleansed with a solution of Permanganate of Potass. and dressed with wet Thiersch dressing for the first twenty-four hours, after which nothing but Bovinine dressings were employed until the patient was discharged with the arm entirely healed, Sept. 20th, 1893. In this case, as in the others, the patient was relieved of all pain after two Bovinine dressings.

CASE No. 2.

NAME.—Sara Lee. AGE.—45 Years. RELATION.—Married. WORK.—Cook. NATIVITY.—United States.

DIAGNOSIS.—Three syphilitic ulcers of the left leg.

HISTORY.—The ulcers had gradually grown larger for the last three years in spite of anti-syphilitic and other treatments employed.

TREATMENT.—Aug. 10th, 1893, the ulcers were scraped and dressed with wet hot Hydrarg. Bichlor. dressings (1-2000) for the first forty-eight hours, after which nothing but Bovinine dressings were used until the ulcers had entirely healed, Sept. 8th, 1893.

Illustrations of ulcers before and after treatment was commenced are on page 29.

CORNEAL ULCER CURED BY TOPICAL NUTRITION.

Dr. Louis B. Couch, of Nyack, N. Y., has described in the *New York Medical Times*, an extreme case of corneal ulceration, which he cured by literal *feeding* of the perishing tissues, with beef serum.

"The central portion of the cornea being farthest away from the source of nutrition, became yellow, infiltrated, and densely opaque, while around it appeared extensive serpiginous ulcerations, which resulted in a serious loss of vital corneal nutritive material. The cornea now became wrinkled, diminished in size, and the eye appeared greatly shrunken. The whole epithelial layer of the cornea also sloughed away, leaving the anterior elastic lamina, dull and lustreless, exposed to view.

"The case was certainly extremely dangerous and apparently hopeless. It was plainly apparent that unless new elements of nutrition were speedily obtained a sloughing of the cornea and an evacuation of the contents of the eyeball were inevitable. I was struck with the idea of direct corneal feeding of beef serum by endosmosis, and immediately proceeded to utilize it."

"Obtaining plenty of juicy round steak, I applied the juices expressed therefrom almost constantly to the wrinkled, shrunken cornea. For the first day I more than held my own, and on the second day there were plain evidences of improvement, which continued uninterruptedly. The cornea gradually lost its haziness, the ulcerations healed without a scar, and to-day the eye is as bright and clear as its fellow. This application of beef serum to the cornea was continued steadily till the inflammation, swelling, and pressure upon the corneal blood-vessels had subsided sufficiently to allow the natural sources of nutrition to resume their wonted functions."

"The absorption of the dense central opacity of the cornea and the reproduction of its epithelial layer required several weeks to accomplish, during which interval I watched and studied with greatest interest the process of Nature building, as it were, a new cornea. Commencing at the inferior sclerotico-corneal border, the new growth crept slowly upward in a nearly straight line, till the whole cornea was plated with a new, beautiful epithelium. When the pupil was partially covered, the affected eye presented a most grotesque appearance, the upper half being lustreless, dull, and dead in appearance, while the lower half was as bright and clear as a polished diamond."

Fig. 5

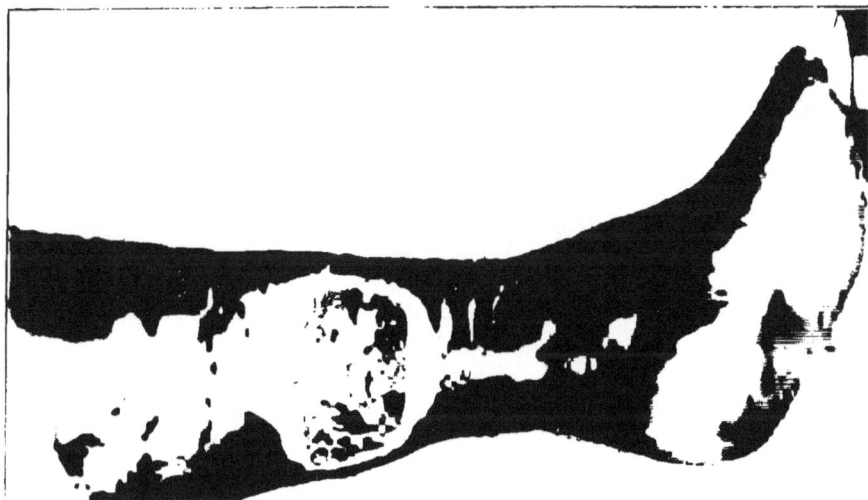

31

[From "*The New York Polyclinic.*"]

MAL-NUTRITION OF BONE, AND REPAIR.

BY

T. J. BIGGS, M. D.

CLINICAL ASSISTANT, DEPARTMENT SURGERY,
NEW YORK POLYCLINIC.

Outline of History and Treatment of an interesting case of Felon.

Ellen Sweeny, twenty-four years old, applied to the New York Polyclinic, October 29th, 1893, seeking treatment for what appeared to be an ordinary felon of the right index finger. The outward manifestations, however, did not determine the extent of the deeper trouble. She gave the history of having pricked her finger with a pin ten days back. This was followed by inflammation and swelling of the finger and surrounding portions of the hand, for which condition she applied for treatment at the New York Hospital, and an incision was made over the seat of trouble. Instead of returning for after treatment and dressing, the patient takes the matter in her own hands and commences to apply the old-time and much abused treatment of poulticing. Not gaining any relief, she came as above mentioned, to the Polyclinic.

An exploratory incision was first made under cocaine to investigate the depth and extent of tissue involved. The result of this exploration revealed a most extensive amount of phlegmonous inflammation to exist, the trouble having evidently started at the flexor tendon.

The primary incision had apparently not sufficiently drained the tissues, and on the top of this, the poultices, encouraging suppuration without drainage, further promoted the progress of disintegration to the detriment of surrounding and deeper tissues, so that at the time of the exploration the flexor tendon for about three inches of its length, was found to consist of one long slough, and in consequence was snapped in half without being submitted to undue tension. The pus had burrowed upward to almost the center of the palmar surface of the hand and downward to almost the distal end of the index finger, being surrounded by an area of broken down sloughing tissue. Furthermore, it had extended in depth to the bone itself, which was found to be denuded of its periosteum for the entire length of the anterior surface of the distal phalanx and part of the second of the index finger. By the aid of a probe, it was determined that the posterior surface was also devoid of periosteal covering, with the exception of a very small area: *thus the bone was denuded for almost its entire circumference.* The general appearance of the surface of the bone which was exposed, presented an almost drab color and did not bleed when the wound was opened, although it was not attempted to do any scraping at that time. The appearance was of such a dark hue and was so devoid of that "pinkish" appearance, which is characteristic of healthy bone, that it was thought there existed but a small chance of saving the finger, and the patient was instructed to return the next day for operation, being advised of the possible necessity of amputation. On the succeeding day, under anæsthesia and with a tourniquet applied to the arm to insure a dry operation, the incision was considerably extended until healthy tissue was reached. The sloughing soft parts were completely scraped away, including almost the entire length of the flexor tendon of the index finger. This resulted in leaving an extensive open wound of about four inches in length and three-quarters of an inch in breadth, leading down to the bone at the bottom, which was exposed through a large portion of the wound, and denuded of its periosteum, presenting on the whole a most unpromising appearance. After scraping the bone only slightly, it was decided to relieve the constriction above and by inspecting determine the amount of circulation still existing in the bony tissue. The general appearance presented was not altered. There was no discernible bleeding of the surface of the bone, and it was a question whether there existed any variation from the general darkish hue which the whole surface presented, and which indicated the almost total lack of nutrition. In isolated spots there seemed to exist minute points, which could possibly be considered slightly pinkish in hue, and it was on this account alone that it was decided not to amputate the finger but to give the patient the benefit of the doubt and do nothing further in the way of operating, after removing all necrosed, soft tissue, and establishing ample and unobstructed drainage. For the first twenty-four hours after the operation the wound was treated antiseptically by the use of wet bichloride ($\frac{1}{3000}$) dressings, and subsequently the only treatment which was pursued and the only application employed was the sterilized liquid preparation known as "Bovinine," consisting of defibrinated ox-blood and desiccated egg albumen, suspended in spiritus frumenti.

The theoretical reasons which dictated the use of this substance, were that a bone so completely denuded of periosteum and deprived of its means of nourishment from without, and showing such incomplete capillary circulation within, would be benefited by the application of artificial nourishment supplied locally.

The patient was seen daily for the first ten days, and later every other day, and each time after the wound had been washed with peroxide of hydrogen, the cavity was filled with "Bovinine," and nothing else used except an outer covering lightly applied. Four or five days after the operation there was a slight change for the better in the appearance of the bone, and from that time on the question of its recovery was no longer in doubt.

The bone and surrounding soft tissues progressively regained their normal appearance, and the whole wound became gradually covered by healthy granulations, the bone itself being the last to become covered in this manner. So soon as it was determined that the whole surface of the wound was completely and entirely protected by healthy granulations, no further attempt was made to keep it open, and strapping was resorted to to bring the edges of the wound together and produce union by granulation tissue.

The process of repair has been uninterrupted from that time on, and at the present day the wound is absolutely healed throughout its entire length. The only difficulty that remains is the loss of flexion of the second and third phalanges on account of the destruction of the flexor tendon.

The points of interest in the consideration of this case are as follows :—

The remarkable tendency of the bone to regenerate after it had been so extensively deprived of its natural means of nourishment and the lack of a necessity for any subsequent operation, there being no superficial exfoliation of necrosed bone. Formerly it was considered that bone denuded of periosteum was beyond hope of recovery, but it has since been demonstrated that while the periosteum is the external nourishing membrane of the bone, its loss is not essentially followed by necrosis, as, in this event, the internal capillaries keep up the nourishment and preserve its vitality. But in this case it will be seen that the capillary circulation seemed very inadequate, and naturally serious doubts were entertained whether there would be sufficient circulation to furnish the required nourishment and restore the bone to its normal condition of health. In cases where this has been accomplished, it has often resulted in only a partial restoration of the bone, and a superficial necrosis has followed, which had to be removed before the wound could be properly closed. In the face of the above unfavorable condition, it was deemed advisable to resort to a means of supplying nourishment artifically by direct applications, in the hope that the loss of nourishment from inadequate capillary circulation might be supplied through an artificial medium by absorption. The result in this case was certainly a most successful one, and its report and consideration may not be without instructive features.—*The New York Polyclinic. (Monthly Journal.)*

New York, November 29th, 1893.

DETAILS OF MODIFIED SKIN-GRAFTING, IN A NUTRITIVE MEDIUM.

F. L. came to the Demilt Dispensary December 22nd, 1892, having sustained two weeks previously a serious injury to his left forearm while pursuing his work as brakeman on the Long Island Railroad, the result of which injury was a large laceration of the forearm, which was denuded of the integument and superficial and deep fascias, to the extent of 7 inches in its long axis and 3½ inches in width, exposing to view a clearly dissected plane of all the superficial layer of muscles on the outer and posterior surface. For the two weeks prior to his appearance at the Demilt Dispensary he had been treated at several other dispensaries and hospitals under antiseptic methods, but in spite of these measures the surface of the wound revealed a large sized black slough, about 2 inches in length and extending down to the periosteum between the supinator longus muscle and the tendon of the biceps. Furthermore, as a result of the severe blow which the muscles of this region had sustained, total paralysis of the forearm existed. The conditions had appeared so unfavorable immediately after the injury that he was advised at the institution he first visited to have the arm amputated. This he refused to submit to ; hence his appearance at the Demilt for other treatment. The preliminary treatment pursued in his case was the employment of such applications as would tend to hasten the removal of the slough and stimulate the wound to a condition of healthy granulation—namely, Balsam of Peru, Peroxide of Hydrogen, etc. The desired condition was obtained in about ten days' time, when it was decided to undertake to heal the wound by skin-grafting. The method employed was not that of Thiersch, but by the direct application of small grafts to the granulating surface. In detail it was as follows :—After carefully cleansing the surface of the wound with Thiersch's solution (corrosive sublimate was not used, as it is believed to coagulate the albumen in the tissues and interfere with the adhesion of the graft), and preparing a clean area on the other arm, from which the grafts were to be taken, several small pieces of skin were removed from this surface by means of a pair of ordinary scissors and dressing forceps, washed with the Thiersch solution and carefully placed over the surface of the wound at a distance of one inch apart each. The cut on page 37 illustrates the wound as it appeared prepared for the skin-grafting. Eight grafts were employed in all. Directly over the grafts were placed strips of rubber tissue, soaked in Thiersch solution, and over this sterilized gauze wet with the same solution. The dressing was completed by the application of more rubber tissue over the whole forearm, and finally a splint and an evenly-applied bandage.

Forty-eight hours after the grafting was performed, when the dressing was removed and the wound inspected, the grafts were barely adherent and had lost their pinkish hue, showing their poor nutrition. They were not disturbed, however, but the character of the dressing was changed, and in place of Thiersch solution Bovinine was employed. From this time on, the grafts were afforded a constant source of artificial nutrition through the agency of the Bovinine, on which pabulum they thrived, developed and became permanently adherent. The cut on page 38 reveals the condition of the wound and the grafts after the Bovinine had been employed several days. The dressings from now on were changed about every forty-eight hours until at the end of about eight weeks, when the wound was found to be absolutely and entirely well, having become covered with healthy epithelium, which had traversed the surface of the wound, having been derived partly from the circumference and partly from the several grafts. The cut on page 39 shows the wound and the grafts at a later stage, when partially healed. The cut on page 40 shows the wound entirely healed.

The noteworthy points in this case were the following :—

1. The rapidity with which the wound acquired a healthy condition prior to the skin-grafting under the use of Bovinine.

Fig. 10

2. The rapidity with which the grafts became adherent to the surface of the wound under the same means. Of all the grafts employed no failure occurred, except that one had been accidentally brushed off.

3. The unusually rapid and complete healing of a wound of such large dimensions.

Finally, it may be stated that those who have attempted to skin-graft by this method and have used other modes of dressing cannot fail to ascribe the complete success of all the grafts in this case, and the rapid recovery of the wound,˙ to the use of a healthy nutritive medium as obtained through the employment of Bovinine.

ANOTHER REMARKABLE CASE.

John Francis, aged 23 years, appeared as a patient at the New York Polyclinic June 2d, suffering from an old burn over the right sterno-cleido-mastoid muscle, of 19 years standing. The general appearance of the wound was extremely unhealthy, and consisted of flabby uneven granulations covered by a sanious discharge. The wound measured about 2½ x 2 inches in dimension, and the granulations, such as they were, overhung and overtopped the wound over one part of its surface, while in another they extended below the edges, covered by a dark greenish foul-smelling slough. In some respects the general appearance presented was that of epithelioma. The patient's general condition was extremely run down and his appearance markedly anæmic. He had lost weight in the last several months to the extent of about 45 pounds. He was unable to attend to his daily work, which was that of a brakeman, on account of the undue suffering caused by the wound, and the pain also deprived him of his night's rest. From the time when the wound first occurred up to the present, he had resorted to every means imaginable to endeavor to produce a healthy process of repair, and had followed the advice of many competent surgeons without success. He had in fact been skin-grafted on two occasions, but this procedure had failed both times. The mode of treatment adopted in this case was the same as suggested in the description of the use of Bovinine in skin-grafting. In this particular case however, the wound was in such an exceptionally unhealthy condition, a longer period was taken for the preparatory treatment, which consisted in daily cleansing, followed by the application of Bovinine as a dressing. By this means the wound was gotten into a sufficiently healthy state to warrant the application of skin-grafts about two weeks after his first appearance at the Polyclinic; a condition which had never before been attained as long as the wound had existed, in spite of the use of all kinds of ointments and astringent and antiseptic washes. On the seventeenth day from the patient's first appearance, three small grafts were taken from previously prepared surfaces, each about one-quarter of an inch in length. The wound by this time, from the treatment adopted, had considerably reduced in size and was perfectly healthy in appearance, but had seemed to have reached a stationary period, having remained exactly the same for about a week, before it was decided to resort to skin grafting. Three grafts were placed in parallel positions on the surface of the wound and held in place by small strips of rubber tissue. The dressing employed over this consisted of sterilized gauze saturated with Thiersch solution, for the first 24 hours after the operation, and following this up to the time of his recovery, which was absolute and complete, Bovinine was the only dressing used. Two weeks after the skin grafting the patient was discharged cured, the whole surface having become covered with healthy epithelium.

SKIN-PROPAGATION FROM CUTICLE SCRAPINGS,

By Means of Topical Nutrition.

The technique of employing skin scrapings to obtain permanent repair of wounds or ulcerations that have obstinately resisted all other measures, is very similar to the method described in skin-grafting. The surface to be treated is previously prepared for a period sufficient to obtain a healthy condition of the wound; this period varies according to the state the wound happened to be in. Having accomplished the above object, a surface of healthy skin or mucous membrane from which the scrapings are to be procured, is properly prepared precisely the same as though a graft were to be taken therefrom. A small dermal curette is employed to obtain the desired scrapings, and with this instrument the surface is *very gently* scraped to remove the exfoliating epithelia, and then once more washed with Thiersch solution, and well dried ; after which the cells to be transplanted on the wound surface are obtained by scraping as in vaccination, without producing bleeding. Enough of these cells are removed to make a small deposit at four or five different points in a wound about two inches in diameter. Following this, strips of rubber tissue are applied and the dressing is the same as in skin-grafting, remembering that after the first twenty-four hours the use of Bovinine alone for the nourishment and support of the new growth of skin, is depended upon to the exclusion of all other dressings. The following are cases in point :—

˙ CASE NO I.

Patrick M. had been coming to the Demilt Clinic for several weeks with a chronic traumatic varicose ulcer of the right leg, over the tibia, which had been treated by ordinary means, at various clinics without avail. After the ulcer had been gotten in a condition of healthy granulation the skin scrapings were procured and deposited on the wound, as above described. The after dressing of the wound consisted entirely of Bovinine. Two weeks after the application of the scrapings the ulcer had entirely healed.

CASE NO. 2.

John Martin came to Demilt Clinic Sept. 3d, 1894, suffering from an indolent ulcer on the right leg, 3 x 3½ inches in diameter. After thorough cleansing of the ulcer and surrounding tissue, a wet Thiersch dressing was applied for the first twenty-four hours, after which the scrapings were applied as described in the above technique and nothing but Bovinine dressings added until the ulcer had healed, twenty-six days after the first dressing.

CASE NO. 3.

Pat. S. came to Demilt Clinic Aug. 16th, 1894, suffering from a varicose ulcer of the leg, 2½ x 3 inches. Patient said the ulcer had not, in spite of all treatment employed, been healed in seven years. The surface of the ulcer was covered with dark, unhealthy granulations exuding a sanguineous muco-purulent discharge. After the surrounding tissues had been thoroughly cleansed, the ulcer was curetted and dressed with bi-chloride of mercury 1-3000 dressing for the first twenty-four hours. At the end of this time the ulcer presented a healthy surface and I decided to try skin-scrapings instead of grafts. Small quantities of the scrapings were deposited on the ulcer at six different points, and dressed as described in the above case. The patient was discharged Sept. 12th with the ulcer completely healed and covered with new skin.

N. B.—In the above described technique the epithelial scrapings may be taken from the palmar surface of the hand or plantar surface of the feet or from the mucous membranes of the external auditory meatus, nasal cavities, or even the buccal mucous membrane. The above technique may be carried out and successfully employed by using instead of the fresh scrapings, small sections procured from the horny growth of an ordinary hard corn, as demonstrated in several cases like the following.

CORN CASE NO. I.

John Polaskie, suffering from an ulceration over the crest of the left tibia about 2½ inches in diameter called upon me, having heard of the Bovinine treatment for ulceration, and, feeling that he had exhausted all other measures, requested that some form of the Bovinine method be applied to his case. I decided to use him as a trial case in the use of skin grafts obtained from the hypertrophied horny tissue of a corn. I treated him for a few days with Bovinine and obtained a state of healthy granulation during which period I applied wet antiseptic dressings to a well developed corn on one of his toes. I then scraped away the outer softened portion of the corn, and then pared off three thin layers of the horny tissue, and placed them at different points on the granulation surface. These were held in place by strips of rubber tissue, and dressed with wet Thiersch for the first twenty-four hours; after which the Bovinine dressing was exclusively employed. On the third day the corn grafts showed that they had taken firm hold, and at the end of third week the ulcer had entirely healed.

CORN CASE NO. 2.

John Skolly (born in Ireland, age 49, married) came to Demilt Surgical Clinic August 8, 1894, suffering from a traumatic ulcer 2½ x 1¾ inches in measure, over the tibia of the left leg. The ulcer was of 15 months standing, and had been treated at four reputable institutions with unsatisfactory results. It was covered with unhealthy granulations, and exuded a very foul muco-purulent discharge. I curetted it thoroughly, removing the unhealthy granulations, and gave it a wet Thiersch dressing for 48 hours. It was then treated with Bovinine for three days as described in former cases; after which, the wound being then found in a healthy condition, five bits of corn shavings about the breadth of a split pea were distributed over the surface, and secured by strips of thin rubber tissue. Over this was placed an ample quantity of wet Thiersch gauze; then a protection of oiled silk, overlapping all two or three inches in all directions; the whole enclosed with cotton and bandaged. After 48 hours this dressing was removed, and all the corn grafts were found feebly adherent. Exclusively Bovinine treatment was then applied from this time on, until, at the end of two weeks, the patient was discharged cured, with only a small, soft, pink cicatrix in the place of the ulcer.

GANGRENE OF THE SCROTUM: THE MEMBER NATURALLY RESTORED!

JOS. L. BLACK, M. D.

COOK COUNTY HOSPITAL, CHICAGO.

William F., age 30, came under my care February 28, giving the following history :

Past health had been very good. During the summer had been serving as a Columbian guard. About three months before his admission to the hospital he contracted gonorrhœa, which assumed a severe type. He was treated with injections and the discharge was materially diminished.

About a month later, on getting out of bed one morning, felt severe pains radiating from the testicles into the groins. These became so intense that he was obliged to return to his bed. The pains continued and a severe epididymo-orchitis developed with quite severe constitutional symptoms.

The circulation in the scrotum was much interfered with and soon the member showed signs of gangrene. This developed rapidly and the line of demarcation soon made its appearance, running completely around the scrotum about half an inch below its origin.

At this time patient came to the hospital in a truly pitiable condition The scrotum was completely gangrenous and almost ready to separate. Its contents were still much swollen and tender. Intense chordee was present and the penis was swollen and extremely painful. A broad region almost across the abdomen, and from near the navel to points several inches down the thighs, was covered with an angry erysipelatous blush. The affected parts gave out a horribly offensive odor and the skin and breath partook of this to some extent.

Patient was feeble and pale. Had frequent chills at irregular intervals. Pulse was feeble and thready, running 110 to the minute;

respirations 30, shallow. Had a profuse, offensive diarrhœa. Temperature fluctuated, usually keeping near 102 degrees.

Treatment.—Was in such poor condition on admission that immediate stimulation was resorted to. When the pulse had been steadied, the affected parts were cleansed, and a boracic acid wet dressing applied. This was continued for 24 hours, at the end of which time the affected skin regained its normal appearance and the scrotum sloughed away, septum and all, leaving the testicles as bare as if they had been carefully dissected out.

Internally, arsenicum 3x,gr.x, and whiskey ℥ ss were given four times a day for five days. Then the whiskey was discontinued and the arsenic given alone for two weeks.

The case was seen by a number of physicians and they were unanimous in the opinion that nothing but a plastic operation would answer, some advising the removal of one testicle.

Instead of resorting to this I resolved to try Dr. Pratt's Bovinine treatment. The testicles were thoroughly cleansed with a solution of boracic acid, then dressed with iodoform gauze saturated with Bovinine and sprinkled with sulphate of quinine. This dressing was changed twice a day. At the end of 48 hours the raw surfaces began to glaze over and, a few days later, granulations sprouted plentifully from the stump of the scrotum. These spread downward and the testicles drew up slightly, and at the end of about three weeks the latter were completely covered in.

Very shortly after the use of Bovinine was commenced, there was a marked improvement in the general condition. Temperature soon remained at the normal point and all other evidences of septicæmia disappeared. The appetite was good, sleep perfect, bowels regular.

At the end of two weeks, patient was able to sit up, and made an uninterrupted progress to recovery. The functional activity of the testicles has been fully regained.

Was discharged April 5th, in perfect health. That this has continued, and that the cure was complete may be inferred from the fact that Mr. F. is now a member of a baseball team and engages in various kinds of active work.

GENERAL TECHNIQUE OF THE VARIOUS METHODS OF EMPLOYING BOVININE.

First.—The internal administration of Bovinine.

Bovinine may be taken internally in doses from five drops to two ounces "pro re nata," combined with either milk or seltzer water or with any other suitable vehicle.

Second.—Rectal injections.

Bovinine may be given by rectum combined with either milk or a neutral salt solution, preferably from four to eight ounces in a preparation of two parts Bovinine to one of the other, and repeated as necessary. An ordinary size soft rubber catheter should be attached to the syringe and introduced about eight or nine inches to reach the sigmoid flexure.

Remarkably successful results have been obtained after severe hæmorrhage and shock, by this method, when the pure ox's blood serves to immediately compensate for the sudden loss of the human fluid.

The manner in which the Bovinine is obtained from the first gush of the carotids simultaneously with the death of the animal excludes the possible presence of any ptomaines, and eminent authorities testify to the absence of any tubercle bacilli.

Third.—Subcutaneous injections.

Bovinine is used subcutaneously where a rapid response is desired, as after a hæmorrhage or severe shock, in a proportion of two parts Bovinine to one of neutral salt solution (teaspoonful plain salt to pint of water) heated to 100 degrees F.

Here, as in rectal alimentation, the wonderful phenomena of osmosis are brought into play, by virtue of which the human fluid extracts from the animal blood what it has lost by hæmorrhage.

In employing this method a good size aspirating needle should be selected and everything should be cleansed and sterilized before proceeding. After carefully preparing the surface, inject the fluid deeply into the soft tissues of either buttock, after which gentle massage should be employed for about ten minutes.

Fourth —In skin grafting.

In this technique the application of Bovinine is resorted to on the second day after the graft has been deposited on the previously prepared surface. It has been practically demonstrated that the use of Bovinine here hastens the adhesion of the grafts, and by furnishing an active supply of nutrition for their sustenance, the success of the procedure is more certainly assured and more rapidly accomplished.

The ·dressings are the same as employed in ulcerations, and should be changed each day and narrow strips of rubber tissue should be laid over the grafts to prevent their being rubbed off.

Fifth.—In all kinds of ulcers.

DIRECTIONS FOR USE.

Having thoroughly cleansed the ulcer and surrounding tissue with warm water and castile soap, apply to ulcer Peroxide of Hydrogen or a five per cent. solution of Permanganate of Potassium. After gently drying the surface with absorbent cotton, the ulcer is ready for the application of Bovinine, which should be made as follows : — Cut four or five pieces of plain sterilized absorbent gauze, saturate them with Bovinine and simply lay upon the ulcer, if it be even with the surface, or if undermined pack beneath its edges ; cover the gauze with sheet lint; and as an outer dressing to prevent evaporation, place oiled silk or muslin (the sheet lint and oiled silk or muslin should be large enough to overlap the edges of the ulcer at least four inches). Over the whole apply a cotton bandage to keep the dressing in position.

The dressings should be changed daily in the early stages, and later every other day.

N. B.—Care should be taken not to apply solutions of Corrosive Sublimate, strong Carbolic Acid, Acetate of Lead, Tannic or Mineral Acids, strong Alcoholic Liquors, Boiling Liquids or any substance which precipitates or coagulates Albumen, co-incident with this method of treatment; or if they *have been* used the surface should be washed with plain water. If antiseptic solutions are absolutely necessary, saturated solutions of Boric Acid, Permanganate of Potash, Peroxide of Hydrogen or Pyrozone (in very indolent cases), may be used, as none of these are capable of disorganizing the Bovinine.

SOME VALUABLE COMBINATIONS OF BOVININE.

In the surgical treatment and dressing of various lesions, highly satisfactory results have been reached by combining Bovinine with the antiseptics or medicaments most appropriate to the case, properly proportioned and chemically compounded. As a beginning, we present the following formulæ, which have been worked out under the direction of medical experience in Hæmatherapy, by an able chemist, and have been successfully tested in practice for the simultaneous and composite action of the ingredients.

I. BOVININE AND MERCURIC CHLORIDE.

Parts:—Bovinine 1,000: Mer. Chloride 1: Sodium Chloride 3.

The Chlorides are first dissolved in water separately: then the solutions are mixed and added to the Bovinine.

In this instance, an albuminate of Mercury is formed, which is dissolved by the excess of Sodium Chloride.

II. BOVININE WITH THYMOL.

Parts:—Bovinine 940: Thymol 25: Alcohol 20: Glycerine 15.

Thymol is dissolved in warm Alcohol: then Glycerine is added: when cold, the Thymol solution is added, little by little, to the Bovinine. By taking half as much again of Glycerine the work is facilitated; but care must still be taken to mingle very slowly, as otherwise a precipitation of albumen will follow.

III. BOVININE WITH IODOFORM 2 PER CENT.

Parts:—Bovinine 100: Iodoform 2: Ether 5: Gum Acacia 5.

In preparing the 2 per cent. Iodoform and Bovinine, little trouble is met with. There is absolutely no change taking place in the Bovinine proper.

IV. BOVININE WITH IODOFORM 4 PER CENT.

Parts:—Bovinine 90: Iodoform 4: Gum Acacia 6.

The 4 per cent. emulsified mixture requires more care. The Iodoform is triturated with ether, the Acacia is added, and an emulsion is prepared with the Bovinine. Two per cent. more of Acacia will form a still better mixture.

- It is worthy of note that in these compounds the obnoxious odor of Iodoform is completely masked.

The facts deduced from a series of experiments have proved Bovinine to be the aliment "par excellence" and to contain every substance necessary for the production of healthy granulations, whether it be implanted in sponge grafts for the cure of osteo-necrosis, applied to muscle-tissue, or for the regeneration of the superficial structure. After a few local applications of this nutritive material there follows a subsidence of the inflammatory conditions, the granulation will proceed to develop with marked rapidity, the secretions will be totally altered in character and diminished in quantity, but will bear a direct relation to the amount of repair being done by the cellular tissues.

Experiments which have been and are now being conducted in the hospitals and dispensaries afford corroborative evidence demonstrating the unprecedented advantages afforded by the above mentioned treatment, not only in chronic, indolent, syphilitic, varicose, and tuberculous ulceration, but also in skin-grafting, sponge-grafting, rectal fissures and ulcerations, osteo-necrosis, deep-seated abscesses, and in those cases where healing must of necessity ensue by secondary intention, or in other words with loss of tissue. In every case thus far reported, uniformly good results have been obtained, and ninety-five per cent. of the more than five hundred cases presented have been positive cures.

Osmosis; Alimentary Osmosis; and Rectal Alimentation.

The theory and practice of clinical nutrition falls short of proper development, if the wonderful physiological faculty of *Osmosis* be not well understood and employed, in a multitude of the most critical situations that confront the physician. This resource overlooked, as it too often is, or the most effective nutrient to be introduced thereby neglected, there is in many cases no other help for the perishing patient. A few elementary memoranda may properly introduce the record of the latest progress in the employment of this great physiological faculty (so to call it) to save life *in extremis*.

Nutriment is passed into the constructive channels and organs of the body not only through the millions of microscopic mouths called absorbents, but also, in no unimportant proportion, through membranes in which there are neither passages nor pores, being tissueless as a waterproof gum. The emulsified food, in this physiological process does not *filter* through the membrane, as if through a strainer, but first gains entrance into the very substance of the membrane by something like a solvent process; which, being of course dependent on some quality in the substance of the membrane itself, is called the power or faculty of *imbibition*. This property operates from both sides of the membrane, bringing both liquids together within it(the blood, and the lymph or chyle of digestion) where the well known law of inter-diffusion of liquids effects a new composition, each side losing something to the other, and partly passing by each other to opposite sides ; but the main body, consisting of the incoming fluid as modified and reinforced by the other, keeps on to its junction with the vital stream. This is *Osmosis :* the exit of the inter-diffused fluids which had met so easily within the membrane by the familiar process of soaking into it from opposite sides; but

53

which offer to our contemplation a mystery in their going forth : a mystery even greater in physiological osmosis than in our extra-physiological experiments ; where we make the membrane of an egg, or a neutral liquid, a diaphragm between two liquids of different densities, or of different composition, in passing through which diaphragm they will so intermingle and interchange as to form one homogeneous solution throughout both receptacles, as if they had been connected by a pipe; or else as in certain cases will make two new solutions differently modified by having exchanged constituents with each other. For, considering the vital results in physiological osmosis, and especially considering the distinct selective powers of different membranes for acquiring (or imparting) widely different secretions from the same nutriment, there is reason to infer a more than mechanical process of entrance and exit—imbibition and osmosis—something more recondite than the imbibitory and osmotic power exhibited in non-vital apparatus.

The employment of the absorptive properties of the mucous membrane of the larger intestine is a useful expedient in all cases where it is desirable to afford the stomach perfect rest or whenever it is impossible to reach the circulation through this organ.

We have in the large intestine a large vascular area which is always ready to receive and imbibe the nutrient material introduced from without for the purpose of restoring and maintaining impaired or lost vitality.

By virtue of the wonderful phenomena of osmosis we are thus enabled to reach the circulation by the use of rectal enemata, and it is a well known fact that a person may be kept alive indefinitely by this means. For this purpose various nutrient substances have been employed as enemata, such as milk, eggs, beef-tea, etc. In making a selection we should choose a preparation that contains all the necessary principles which sustain life, is readily assimilable and unirritating. After careful experimentation I find that Bovinine fulfills these requirements in a most satisfactory manner and is a valuable resource in all cases where rectal feeding is relied upon to support life. A still more important factor is the mucous membrane of the large intestine, as a means of directly reaching the circulation, where prompt action is imperative, as after sudden and severe hæmorrhage. Here the system is suffering a deprivation of its most vital properties and often if restoration is not immediate the loss becomes irreparable. Under these circumstances it is essential to have at hand a preparation which can immediately supply the loss the system has sustained, and this is no doubt the explanation of the great success attending the use of Bovinine under these circumstances, as will be seen in the subsequent report of cases.

No one doubts that in cases suffering from Cancer or Ulcer of the Stomach, Gastritis or Stricture of the Œsophagus, persistent vomiting, shock, or hæmorrhage, that rectal alimentation is very desirable, and that with colonic feeding we can get better and more prompt results, when we are using non-irritating enemas, than we can by any other means.

Formerly rectal alimentation was not often resorted to, because patients were unable to retain it for any length of time on account of the irritation produced thereby. There seems to be no reason to believe that any special secondary digestive process occurs in the cæcum or any part of the large intestine, but it is possible that constituents of the food which has been partially digested and escaped absorption in the small bowel, may become absorbed and assimilated in the large intestine; and the power of this part of the intestinal canal to digest fatty, albuminous or other matter, may be gathered from the good effects of nutrient enemata.

Probably the quickest and best method that can be used in case of emergency is to mix,—

Bovinine..........1 oz.   or   Bovinine..........1 oz.
Milk..............4 oz.   Water............4 oz.

Pancreatine or Chloride of Sodium may be added to either of the above formulæ   Or, the following formula may be used to better advantage:—

Powdered Starch..........................3 drachms.
Cold Water...............................½ oz.

mix and pour into five ounces of boiling water and cook thoroughly. When cooled to 115 degrees F., add,—

Forbes' Diastase..........................30 minims.

Digest for 15 minutes, strain and add
Bovinine...................................1 ounce.
To be used for each injection.

Directions for use:—Place the patient on the left side, or knee-chest position, and carefully introduce a well-oiled tube, connected with a syringe filled with the fluid. Having inserted the tube, gently force it along, at the same time pumping in fluid to distend the bowel after the sigmoid flexure is reached. Some difficulty may be experienced in passing the sigmoid flexure, but by care, patience and perseverance, it will be easily overcome.

If, in spite of the patient's efforts, the bowels must be emptied, allow them to do so and again introduce the tube and give another injection.

Injections should not be used warmer than 105 degrees, or colder than 80 degrees F.

Huber has recently shown by actual experimentation, that from 58 per cent to 70 per cent of fluid egg albumen may be absorbed from the rectum without peptonization. A slightly larger proportion of albumen was absorbed after peptonization, but less than half as much when chloride of sodium was not added. The proportion of salt found necessary to stimulate absorption, was one gram, or one-fourth of a dram, for each egg.

CASE No. 1.

Mr. B. came to Dr. K's private hospital May 2nd, 1894. An examination revealed an atonied bladder with a rigid neck, for which a perineal section was performed. At the time of operation the patient had about two per cent. of albumen in his urine, and undoubtedly pyonephrosis. During operation patient bled only slightly, but six hours after, packing of perineal wound was resorted to, to stop apparent oozing from wound. On the second day after operation the patient had a violent spasm, followed by copious discharges of blood through the perineal tube. Pressure at first seemed to control this unexpected complication, but only to be followed by a recurrence a few hours later. This state of affairs continued for several days in spite of all measures to overcome it, such as pressure, hypodermic injections of ergotin, &c. After the patient had bled almost constantly for one week, his condition became so alarming that recovery was almost despaired of. The bleeding was finally controlled by the use of alum solutions thrown into the bladder. In spite of the cessation of hæmorrhage the patient's exsanguinated condition presented small hopes of recovery. It was at this time that the rectal enemata of Bovinine were resorted to, in the following manner. A catheter was introduced into the rectum up to the sigmoid flexure of the colon, and six ounces of Bovinine and salt water injected into the rectum, (the injection was one-fourth salt water and three-fourths Bovinine). As the patient was unable to retain anything by the stomach he had to depend entirely on this means of nutrition, as a result of which he made a rapid and uninterrupted recovery. An examination of his urine just before he left the hospital revealed the presence of only about one-eighth of one per cent. albumen against two per cent. prior to the operation. On leaving the hospital the patient was instructed to take one ounce of Bovinine in milk every three hours. He is at present enjoying health, having entirely recovered from the shock of the operation and hæmorrhage.   T. J. BIGGS, M. D.

RECTAL ALIMENTATION.
A DESPERATE CASE WITH GRATIFYING RESULTS.

CASE No. 2.

Miss B., age 16, of Lincoln, Neb., was admitted to hospital in Kansas City, Mo., June 9th, 1891. Laparotomy for ovarian cyst performed on June 12th. History of the case in brief as follows :

Anæmic in the extreme when admitted and generally in bad condition for an operation, but the case demanded immediate relief and the operation was deemed particularly successful, but the low vitality and extreme nervous irritability of the patient gave no promise of a favorable outcome.

Shortly after the operation the stomach became so irritable that all nourishment and even cold water were rejected. The temperature and other grave symptoms indicated sepsis. On June 18th, the date of my first visit to the hospital, the patient's life was despaired of, and the last rites of the church were being administered at the time of my arrival. Dr. G., the surgeon in charge, kindly gave me a history of the case and I at once suggested rectal feeding. This had already been tried with unsatisfactory results, beef tea and milk having been used. At my earnest request I was permitted to test the value of Bovinine, the doctor saying at the time the patient would not live 48 hours. Bovinine, one ounce, sterilized water, one ounce, Pancreatine, five grains, raised to a temperature of 100 degrees F. were employed and forced high up into the rectum This was retained and the same dose was repeated after an interval of two hours. After eight hours the distress and painful retching subsided and if food was not alluded to the stomach remained tranquil. For twelve days the only nourishment administered was Bovinine every three hours day and night, and by this process of nutrition alone, the vitality of the patient was restored so that at the end of that period she sat up in bed and, for the first time since the operation, expressed a wish for food.

Bovinine with carbonated water, followed later with Bovinine and a little diluted cream were given. From this time rectal feeding was discontinued and on July 3d, this moribund girl was pronounced convalescent.

The peculiar value of Bovinine as Rectal Aliment lies in the fact, first, of its ready adaptability to the wants of the system and because all unassimilable matter has been eliminated, thereby avoiding rectal irritation. Thus this product containing as it does all the nitrogenous elements necessary to sustain life, is at once appropriated and used in the repair of the waste going on in the organism.

W. H. PARSONS, M. D.
Omaha, Neb.

CASE No. 3.

Bedford, Pa., Aug. 31, 1892.

DEAR SIR:—Your representative kindly left with me last spring some samples of your Bovinine, and now I remember that I owe you a debt of gratitude which I cannot sufficiently give expression to; not for the samples, but for placing within my reach and that of the profession, so valuable a remedy as Bovinine.

Its stimulant and nutritive qualities are seen in all cases where it is given, but I wish to speak particularly of a case that came under my notice last summer. A Mr. B. is a railroad official high up in the service of the N. Y. Central R. R.

He came to Bedford early in the summer of 1891; his case was thought to be a hopeless one and was diagnosed by one of the most prominent physicians in New York City, to be cancer of the stomach. We kept him alive for days at a time by rectal injections of Bovinine, and when the stomach would bear *anything* administered Bovinine by the mouth, until he was built up sufficiently to take solid or semi-solid food, with Bovinine as a support between other forms of food, and last fall he was so far recovered as to be able to resume his duties as railroad manager, and this summer he is still improving and I feel confident that if he had not had such support as was afforded by Bovinine, he could not have gone through safely, and I meet people at our Spring Hotels nearly every day, who are loud in their praise of Bovinine for bringing themselves and friends safely through some trying ordeal of sickness, in connection with the best remedies used.　　　　　I am, Respectfully yours,

J. A. CLARK, M. D.

CASE No. 4.

514 Colerain Ave., Cincinnati, Ohio,
April 24, 1893.

To the Bovinine Co., New York.

GENTLEMEN:—I am glad to give my experience with Bovinine in rectal feeding in a severe case of vomiting in pregnancy a few months ago. At the suggestion of one of your representatives, I used a formula consisting of half teacup thick boiled starch, digested with Forbes' Diastase, to which was added one tablespoonful Bovinine. This quantity was given every three or four hours for a period of three weeks, the patient taking nothing by mouth, except occasionally a little water. The rectum was washed out twice daily with salt water before the administration of the nutrient enema. A rectal tube was not used, all of the injections being made by an ordinary syringe; still, the rectum was not in the least irritated, and I believe not a single injection was expelled. The patient gained considerably in appearance of weight, very much in strength, becoming able to sit up in a chair, notwithstanding that she was unable to raise her head off the pillow when treatment was begun.

Very truly,
W. E. SHAW, M. D.

SUBCUTANEOUS AND RECTAL INJECTION OF BOVININE.

The following cases which occurred under the personal observation of Dr. T. J. Biggs contribute further evidence of the rejuvenating properties of Bovinine as used by rectal *and subcutaneous* injection after severe hæmorrhage.

September 19th, 1894, I was called to see Mr. V. who had lost a large quantity of blood, the result of several incised wounds received in a fight ; one cut had opened the radial artery. The hæmorrhage had been controlled prior to my arrival, but the patient had lost so much blood that he was unconscious and his life despaired of by the family physician and friends. I immediately stimulated the patient with hypodermic injections of Strych.-Sulph., Glonoin and Tr. Strophanthus, following which ten ounces of Bovinine and salt water were injected subcutaneously, four into the right nates, four into the left nates, and two into the right thigh. Twenty-three minutes after the last injection the patient regained consciousness with a decided improvement of the pulse and respiration. A rectal injection of five ounces of Bovinine and salt water was now employed, and an hour after, the patient fell asleep, from which he awoke one hour and thirty-five minutes later, much improved. For the following three days rectal injections of Bovinine and salt water (six ounces each) were administered every three hours. The patient has improved daily up to date and is now out of danger, with the wounds healing kindly.

T. J. BIGGS, M. D.

AN EXTREME CASE OF POST-PARTUM HÆMORRHAGE.

June 23d, 1894, I saw with Dr. S., Mrs. T. who had lost a large quantity of blood, the result of a post-partum hæmorrhage. The patient was in a collapsed state and presented all the symptoms of that form of anæmia following the loss of large quantities of blood. Her pulse was very weak and ranged from 173 to 179, the skin was cold and clammy and the respiration rapid. It was decided to stimulate her with hypodermics of Strych.-Sulph. and Glonoin, and then employ subcutaneous injections of Bovinine and salt water. One twentieth grain of Sulphate of Strychnine, one fiftieth grain of Glonoin and four drops of Tr. Strophanthus were injected, and immediately following this six ounces of Bovinine and salt water were injected into the right nates, and four into the left, with the most happy results.

In an hour the patient's pulse had dropped to 99 and was stronger ; her respiration was almost normal. A rectal injection of six ounces of Bovinine and salt water was now administered, which the patient retained and in about an hour she fell into a quiet sleep. For the following three days the patient received nothing but Bovinine, milk and sherry wine.

Two weeks from the date of the first subcutaneous injection the patient was in full possession of her health.

T. J. BIGGS, M. D.

A VALUABLE AGENT IN THE TREATMENT FOLLOW-
ING PARTAL AND POST-PARTUM HÆMORRHAGE.

By Dr. STAFFARD.

It is a well known fact that a patient's life is still endangered after the loss of a large amount of blood either at the time of labor or from a post-partal hæmorrhage, even though the temporary reaction seems fairly good. Should the body not be made to rapidly replenish itself with vital principles similar to those lost by the hæmorrhage, there is left a weakened condition the ready prey of disease which in a period varying from a few months to a year or more will probably manifest itself in some form. Hence, it should be our aim to restore the lost vitality as early and as effectually as possible. The nutritives used in these cases are numerous. They consist of eggs and various preparations from milk, beef, blood, etc. All of these preparations are nutritious and valuable, but vary to quite a degree as to their digestibility and assimilability in conditions of low vitality. To over-come past difficulties experts have for years been vieing with one another to form a combination of nutritives which would contain the largest amount of nutritive material and be at the same time the most readily digested and assimilated. The best preparation now on the market is Bovinine. It is prepared by The Bovinine Co., of London, New York and Chicago, and reflects great credit upon the manufacturers, as more nearly meeting the indications in these anæmic cases than any of the many prepared foods. Bovinine is not a new preparation, having been upon the market about fifteen years, but its value as a reconstructant, especially in labor cases attended by hæmorrhage, undoubtedly has not been fully appreciated by the obstetrician. For those unfamiliar with this preparation it is perhaps well to state that it is a dark brown fluid consisting of defibrinated beef blood and other vital principles obtained from beef by a cold process of extraction, desiccated egg albumen, whiskey in sufficient quantity to act as a preservative, and a few other ingredients which act as palliatives and favor absorption. The process of extracting the vital principles from beef without the aid of heat is undoubtedly a very good one and probably furnishes more of the nutritive material contained in the beef and in a manner to be more readily cared for, than by processes in which heat is employed, as the effect of high temperature upon albumen and blood corpuscles is known to be unfavorable to the ready digestion of the former and to destroy the latter. A drop of Bovinine examined with the microscope reveals the presence of a substance containing a large number of healthy red blood corpuscles. Knowing in a general way the composition and process of preparation of Bovinine, it is readily seen that two things should be avoided in its administration, the one is heat and the other any strong preparation of alcohol which would produce a coagulation of the albumens and destroy the blood corpuscles.

Another good point about this preparation is the tolerance which an irritable stomach has for it. If any food will be retained Bovinine is that food. Should the stomach refuse to tolerate food we have a useful remedy in Cocaine Hydrochlorate, of which a drop of the four per cent. solution should be given in a small teaspoonful of Vichy every ten minutes for three or four doses, after which the Bovinine will usually be tolerated in small amounts. Bovinine when administered by the mouth probably furnishes the best results when combined with peptonized milk. At first if the patient is very weak it is better to begin with one or two drachms of Bovinine to two ounces of peptonized milk, to which may be safely added a small amount of brandy or whiskey ; but if the conditions require any degree of stimulation it would be better to administer the stimulants at such times when they will not come in contact with the Bovinine in the stomach. The smaller quantity of Bovinine and milk may be repeated every half hour, and if well tolerated may be increased so that from one and one-half to two ounces of Bovinine may be administered in eight ounces of peptonized milk every three or four hours. It will be surprising to the obstetrician who uses this preparation for the first time to observe the more rapid improvement in his patient than he has been accustomed to see by the employment of other prepared foods. The improvement may be still more rapidly "pushed" by administering Bovinine by the rectum, as it is usually well tolerated and quite readily absorbed from the lower bowel. It should be given in from two to three ounce quantities diluted with about a third of its quantity of water to which a small amount of table salt has been added. The enema should be given warm, high up through a rectal feeding tube, care being observed to inject the solution slowly, thus allowing the bowel to acquire a greater tolerance to its reception.

These enemata may be repeated usually three or four times in the twenty-four hours, and in especially weak cases may be kept up for some days, care being observed to use laxative enemata occasionally so as to leave the gut in a favorable condition for absorption.

The above described plan of treatment is one which should appeal to all physicians doing obstetric work, and if once employed will be found to be several paces in advance of other methods heretofore used in this class of cases.

The following is a report of three cases treated by Bovinine.

CASE 1.

Mrs. K., age 21, primapara, family history bad, having lost both parents and a brother, of phthisis. The patient was poorly nourished, anæmic, not well developed, having a diminished chest expansion, and had scarcely seen a day when she felt perfectly well, from childhood to the time of taking this history. Sudden changes in temperature had an unfavorable effect upon her, so that for the past four or five winters she has had a "cold" which lasted the greater part of the season.

During early life she was taken South each Winter, but for the past five years has lived north. As puberty was reached dysmenorrhœa and catarrhal endometritis added to her discomfort. Later on she was advised to marry, but this change seemed to have but little effect other than to lessen the severity of dysmenorrhœa.

After marriage about four months she became pregnant, but even this change seemed to have but little effect upon her general nutrition. The patient went to term with but little change ; the labor was tedious because of the inefficiency of the pains, and as the patient was getting weak and the friends anxious about her, the writer was called in consultation. The usual low-forceps operation was performed, and the placenta removed with but little attendant hæmorrhage, but it was noticed that the uterine contraction was inefficient. About two hours later the writer was again called, to find the patient having quite a severe hæmorrhage, but one which was readily controlled by ergotol and hot intra-uterine and vaginal douches. During the hæmorrhage the pulse became so weak it was difficult to distinguish the beats, the surface became cold, and the patient was gasping for breath. In addition to the usual treatment the writer advised a high warm enema of Bovinine ℥ i ss, Spts. Frumenti ℥ ii, Aqua Fervens ℥ i, table salt q.s. to be repeated every four hours, and by the mouth Bovinine ℥ iv and peptonized milk ℥ ii every half hour. ,The patient showed improvement as time elapsed; the enemata were well absorbed, as was the nourishment taken by the stomach. The enemata were continued after the first thirty-six hours only three times in the twenty-four hours for a week, after which they were discontinued. The physician in charge of the case at first objected to this plan of nourishment, but seeing the marked and rapid improvement in the case, said he believed with the writer, that it gave the best and quickest results he had seen. Bovinine was continued by the mouth, ℥ ii three times a day for about four months, at the end of which time it was no flattery to say she looked better, felt stronger and weighed more than ever before.

CASE No. 2.

Mrs. W. L., aged 28, II-para, family history good. Her previous labors had been prolonged, but otherwise normal in every way. The present labor was complicated by an accidental hæmorrhage, necessitating rapid delivery. The large quantity of blood lost produced the usual symptoms in exsanguination, syncope, cold surface, feeble pulse, etc. The usual treatment, such as surrounding the patient with hot bottles, elevating the foot of the bed, administering hypodermic injections of ergotol, strychnia and brandy, was done. A high enema of Bovinine ℥ ii, warm water ℥ i, table salt qs., was given and repeated every three hours for three doses, after which it was given only four and later three times in the twenty-four hours for ten days. Small quantities of Bovinine and milk were given by the mouth every half hour, and as more could be taken, were increased to two ounces in a glass of milk every three or four hours and later on given three times a day. This treatment was continued four months, at the end of which time the patient considered herself well.

CASE 3.

Mrs. Annie T., age 29, VI-para, family history good. She had always been in good health with the exception of having the diseases of childhood, up to the time of her marriage. She gave birth to six children in nine years and nearly always became pregnant while she was nursing, thus really never giving the uterus a chance to undergo complete involution from one delivery to another pregnancy. The patient was of good physical development, had a roomy pelvis and her labors had been not at all difficult.

64

At the last labor, which occurred on Oct. 17, '93, upon vaginal examination there appeared to be no abnormality, but a short while subsequent to the commencement of the real labor pains and dilatation of the cervix uteri there was a profuse hæmorrhage which was found to be due to an undiscovered placenta previa marginalis. Delivery was hastened as much as possible, but not until exhaustion was quite marked. The pulse became very weak and the advisability of transfusion considered, but conveniences were not at hand to do it, and as there was a disagreement between the consultants it was abandoned. The heart was stimulated by Strychnine and whiskey. Bovinine ℥ ii, warm water ℥ i, table salt q. s. was given by high enema and repeated in four hours. In the meantime Bovinine and peptonized milk was given by the mouth, alternating with brandy and digitalis.

The reaction in this case was slower than in either of the former two reported, but as soon as it had taken place sufficiently to favor absorption the improvement became more marked, so that at the present time she is feeling quite strong, but has not regained her former flesh and strength.

SOME CASES OF OTHER PHYSICIANS.

Lewiston, Me., April 12th, 1887.

GENTLEMEN;—Permit me to speak in the highest praise of your Bovinine. I have tested it in several cases, notably in one severe case of Placenta Previa, which occurred two years ago in one of the most dangerous of all the presentations in child-birth, my experience with which I will relate. I had just returned from the Philadelphia Lying-in Hospital, where I had been for the Winter for study in the department of Obstetrics, and the lady had been awaiting my return for some weeks before her confinement two years ago. She had had several severe hæmorrhages, which had weakened her very much, and she was in a critical condition to undergo parturition. On my arrival I was called to her bedside, and on examination I found her very weak and exhausted. There was no time to lose and I stimulated with Bovinine and brandy. I immediately ruptured the membrane, and turned and delivered the child as soon as the os uteri could be dilated sufficiently.

From the great loss of blood both before and after delivery, the case looked doubtful. She could retain nothing but Bovinine, with a little brandy and milk, for weeks. I am confident that Bovinine saved her life, and to your splendid fluid food both this mother and child are indebted for their existence, as the child was violently ill a year after from cholera infantum, and was brought through by

65

Bovinine alone, the great nutritious qualities of which are a boon to humanity. I have used it in many cases and find it always to be relied upon. You are at liberty to use this letter in the interests of suffering humanity.   W. S. HOWE, M. D., 135 Ash Street.

Toronto, Ontario, Nov. 1st, 1887.

GENTLEMEN:—I have used Bovinine with great satisfaction in several cases where a highly concentrated nutrient was indicated, and especially in partial collapse from hæmoptysis. I have found that it caused systemic reaction with almost the rapidity of transfusion. In all cases of severe hæmorrhage from any cause, or in low and depleted conditions of the body resulting from typhoid fever or septicæmia, and more especially in diphtheria, gastritis, gastric ulcer or malignant disease of whatsoever nature, its place cannot be supplied by any other preparation.   R. HEARN, M. D. C. M.

St. John, N. B., March 3rd, 1887.

GENTLEMEN:—I have used large quantities of Bovinine in my practice during a year past, and have found it especially useful in restoring the strength after excessive uterine hæmorrhages and in irritable conditions of the stomach, as well as in all cases of exhaustion or debility from whatever cause.

GEORGE A. HETHERINGTON, M. D.

BOVININE IN RECTAL INCRETION.

Joseph M. Matthews, M. D., Professor of Principles and Practice of Surgery, and Clinical Lecturer on Diseases of the Rectum at the Kentucky School of Medicine, said: "I cannot understand why anyone would advise Colotomy in cases of ulceration, per se, of the rectum. With strict antiseptic precautions the rectum can be kept perfectly clean, and by the aid of the different meat extracts and fluid foods in the market the bowel can be absolutely rested any length of time.

"The milk diet, as recommended by Mr. Allingham, can be used, or what is better than all in my experience, the preparation called Bovinine, which contains 26.58 of soluble albuminoids, and is the vital principle of beef obtained by a new process. It is a raw fluid food extract, and admirably suited to those cases which require, during treatment, entire abstinence from the solids. I have kept patients for weeks on this preparation alone, during which time local applications were made to the bowel until all ulceration had healed."

Bovinine as a Food Stuff in Typhoid and Gastro-Intestinal Diseases.

By Dr. MAC GRAFF.

Bovinine is a concentrated, easily assimilable, nitrogenous food stuff, especially adapted to the treatment of those diseases accompanied by marked wasting of the tissues; particularly diseases of the alimentary tract and long-continued fevers.

In typhoid fever, a disease in which the waste of the nitrogenous elements of the body is excessive and where the function of assimilation is markedly impaired, Bovinine is an ideal food and is of especial value, being concentrated, and very readily absorbed for the most part from the mucous membrane of the stomach. Not only is the excessive emaciation absent, but during the whole course of the disease the heart action remains much stronger, and owing to the prevention of the excessive waste and degeneration of the tissues, there is not the usual need for cardiac stimulants.

It is best given in quantities of from two to four teaspoonfuls in each glass of milk.

In acute gastritis where the irritability of the mucous membrane of the stomach is marked and all other food is rejected Bovinine in small quantities will be retained, and being easily absorbed, allows the inflamed mucous membrane to regain its normal tone.

In the treatment of chronic gastro-intestinal diseases, from its concentration and readiness of assimilation it is most valuable. In bad cases of gastric dyspepsia where the emaciation and weakness are the most marked symptoms, the administration of Bovinine gives results that are most satisfactory. In the worst cases it may be necessary to give Bovinine exclusively for a few days, or combined with milk; with milk it is given in quantities of from one to three teaspoonsful three or four times a day.

In other cases the simple regulation of the diet, chiefly milk and Bovinine, and proper mode of life, are all that are necessary to effect a cure.

In these cases the effect may be noticed from the start and is continuous and steady.

In anæmia, especially the cases with emaciation and marked gastric symptoms and defective assimilation, the use of Bovinine produces the most satisfactory results. The Bovinine is given, a teaspoonful before meals in three times the quantity of milk or water.

In carcinoma of the stomach where all other foods are vomited, Bovinine is retained and actually prolongs life, preventing the rapid emaciation and starvation and the pain and discomfort caused by ingestion of less concentrated food.

In acute and chronic inflammation of the intestinal tract accompanied by emaciation and diarrhœa Bovinine is again a valuable agent—many obstinate cases of chronic enteritis and colitis being cured by a diet exclusively of milk and Bovinine.

In treating the gastro-intestinal diseases of children Bovinine proves itself to be very useful—many cases of cholera infantum which reject every food, will retain Bovinine given in small quantities at frequent intervals—many of these cases if the nutrition can be kept up recover, that would otherwise die.

SPECIMEN CASES IN TYPHOID FEVER, ETC.

Indianapolis., Ind. Dec. 30th, 1887.

GENTLEMEN:—I have had the pleasure of using your fluid food, Bovinine. I have used it in several cases of typhoid fever and have observed that it sustains the patient's strength better than broths, beef-teas, etc., and is better borne by the stomach, being less bulky. It hastens convalescence and enables them to better resist the disease. The improvement from its use is noticeable to the patient himself and relations. Its effects are so gratifying that the patient asks to be supplied with it. I have used it in pneumonia, where its effects are equally well marked. I have also used it in a case of pyonephrotic abscess, the patient subsequently undergoing an external urethrotomy complicated by a perineal abscess, a fistulous tract connecting abscess with urethra. The patient, debilitated by the pyonephrotic suppuration, was not well prepared for the subsequent operation and disease. The tissues wasted and his vitality diminished. He became so weak that his relations became alarmed. It seemed impossible to secure union. Bovinine was administered at this time and union dated from the commencement of its use. The patient soon was strong enough to sit up. He is improving rapidly under its use. Care should be exercised in beginning its use, commencing with a small dose. It gives me pleasure to say that I cannot speak too highly of its use.

G. W. COMBS, M. D., 22 E. Ohio St.

Pittsburgh, Pa., Dec. 10th, 1887.

GENTLEMEN:—I recently made use of your fluid food Bovinine, in the case of my own wife, who was suffering from debility brought on by lung trouble. She was utterly unable to take any food at all, as her stomach was in so weak a condition that it would reject everything. The Bovinine, though, was retained, and she lived for upwards of two weeks on Bovinine solely. The fluid was administered at first, a teaspoonful every three hours, and gradually increased to a tablespoonful four times a day. The patient gathered strength under this treatment very rapidly, and now, after three weeks' time, is able to take solid food and is rapidly regaining her normal health.

I have also used Bovinine in a variety of cases of typhoid fever with most excellent results, and am now using and recommending it in every case where there is nourishment necessary. It certainly is the highest and best form of concentrated nourishment of its kind that ever was brought to my notice, and I take great pleasure to call the attention of my professional brethren thereto.

Respectfully yours, JOS. B. ENOS, M. D.

Cincinnati, Ohio, Dec. 27th, 1887.

GENTLEMEN:—I take pleasure in saying that in a practice of twenty-two years I have never used a beef food that has given me the satisfaction that your Bovinine has. Samples of it were given me by Health Officer Dr. Byron Stanton. I commenced the use of it on several patients under my care, who were typical cases of typhoid fever, and after the second or third day I noticed great improvement in the pulse. Their stomachs retained it without the least appearance of nausea. It is entirely free from unpleasant taste or smell. I am certainly well pleased with it, and as a new and vitalized blood-maker, am sure it is all you claim for it.

H. H. WATSON, M. D., 399 John St.,
Assistant Health Officer.

Chicago, Ill., Nov. 3rd, 1887.

GENTLEMEN:—I have been prescribing Bovinine in hospital and private practice for the past two or three years in cases of mal-nutrition or wasting produced by typhoid fever, tuberculosis and allied conditions, and find it of marked benefit in sustaining the strength of the patient. I usually combine it with milk.

D. A. K. STEELE, M. D., 1801 State St.

[President of the Chicago Medical Society and Professor in the College of Physicians and Surgeons.]

Winona, Minn., Feb. 4th, 1888.

GENTLEMEN:—I have used Bovinine with great satisfaction in cases of mal-assimilation, chorea, and typhoid fever. In the latter disease it is a nutrient of wonderful value. In one case, a young lady whose temperature reached 105° for two days, with other very unfavorable symptoms, it did glorious work and it was continued during subsequent convalesence.   Very truly,

L. G. WILBERTON, M. D.

Indianapolis, Ind., Dec. 30th, 1887.

GENTLEMEN:—I have given your preparation, Bovinine, a thorough trial with convalescents from typhoid fevers and gastric troubles with a great deal of satisfaction. I have also used it in enteric diseases of children with wonderful results. I have truly found the preparation quite as valuable it was represented to me.

F. O. CHAMBERS, M. D.

Detroit, Mich., Jan. 10th, 1888.

GENTLEMEN:—I have used your Bovinine in several cases of typhoid fever with the most satisfactory results. Have used it in two cases of persistent vomiting in pregnancy, also with excellent results.   F. D. WHEELER, M. D., 452 Antoine St.

Indianapolis, Ind., Dec. 8th. 1887.

GENTLEMEN:—I have used your Bovinine in a case of typhoid fever with great effect. The patient's assimilative powers were restored at once, she being almost moribund.

W. F. BARNES, M. D.

[Expert in Dis. of Mind and Nervous Diseases.]

[From the MEDICAL TIMES AND REGISTER, Dec. 24, 1892.]

THE DIET IN TYPHOID FEVER.

BY

WILLIAM F. WAUGH, M. D.

According to Hoffman,[1] the salivary glands in typhoid fever exhibit almost constant alterations. In the first week they are denser and firmer, of a brownish yellow color and tense in feel. The septa are like cartilage, and creak on section. The acini are filled with closely crowded, very large, multi-nuclear, granular cells. Later on, numerous fatty granules appear in these cells; they become turbid, lose their sharp outlines, and a part breaks down. The glands then become redder and softer, and the tension diminishes. It seems that we have here a parenchymatous degeneration, resembling analogous processes in other organs except that an increase of the cells precedes the degeneration.

These changes, in which all the salivary glands participate, explain the scantiness of the saliva. The pancreas also exhibits perfectly analogous changes.

The secretion of bile is often markedly diminished. The hepatic cells become granular, the nuclei disappear, and fat globules are visible. In bad cases the cells break down into granular, amorphous detritus. In 174 typhoid livers Hoffman found only thirty-eight presenting little or no change; while all the others showed marked degeneration and destruction of the cells, or marked new growth. The secretion of bile is often quite scanty, as may be inferred.

In the intestines the affection of Peyer's patches and of the solitary glands is acknowledged to be the characteristic primary lesion of the disease.

It is necessary also to remark that all the glands are not affected by the morbid process. In a large number of cases the solitary follicles entirely escape; in many others those of the ileum are alone attacked, and when those of the large bowel suffer it is often only in the coecum or in the ascending colon. The same thing is observed of Peyer's patches, the highest of them are very seldom, if ever affected.[2] No special lesion of the stomach is described by Lieber-

---

[1] Ziemssen's Cyclopedia, vol. I, p. 113.
[2] Fagge's Practice, vol. I, p. 200.

meister, Strumpell or Fagge. Wilson says that this organ is in many cases healthy, but occasionally it is the seat of hyperæmia, softening and superficial erosions of the mucous membrane. [3]The secretion of the gastric juice is, however, partially or wholly suspended, as is the case with the otner digestive secretions.

The condition of the digestive system during an attack of typhoid fever, as shown in this brief sketch, warrants our first proposition: That during the course of this fever the power of digesting food is impaired, always seriously, and sometimes almost entirely lost, from the suspension of secretion.

My second proposition is, that food that will not be digested in the stomach or bowels of a typhoid fever patient, is not only useless but harmful; as in the absence of digestion, decomposition is certain to occur, with the production of substances that are certain to be injurious to the patient. I must refer to this cause the tympanites that occasion so much trouble in improperly fed cases. I am assured by the results of numerous observations that this symptom does not become prominent, or even noticeable, when the diet is properly guarded.

The conclusion is, that in typhoid fever the stomach and bowels should not be looked upon as digestive organs, but simply as recepticles for food that has been previously digested.

That absorption may take place cannot be denied. The whole gastro-intestinal mucous membrane is adapted for absorption. Saline solutions, grape sugar and peptones, are absorbed from the stomach. The most active area for absorption is the upper half of the small intestine—the section least affected in typhoid fever. While this absorption is diminished by the presence of catarrh, it is not entirely lost. But as the catarrh, as well as the graver affection of the intestinal glands, become more marked as we proceed downwards from the stomach, the indication becomes very clear: that in typhoid fever we should employ such food-principles as are absorbed from the stomach and the upper bowel. Our list, then, of available foods, comprises water, salts, peptones, maltose and dextrose; while casein, egg-albumen, dextrin and gelatine are not absorbed with the same readiness. As fatty substances are principally absorbed by the lacteals, but little of these can find entrance to the system.

We now come to consider the application of these principles in practice. In France, the obstacles to digestion and absorption appear to be exaggerated. Assuming that only salts and water can be absorbed, and that the patient practically lives upon his tissues, Dujardin-Beaumetz returns to the Hippocratic diet, of water-soup.

3 The Continued Fevers, p. 207.

Liebermeister insists on the importance of giving abundance of water. He objects to the proteids, and falls back upon the carbohydrates, barley-water, oat-meal, gruel, and weak meat-broth. Milk may be given, if boiled and diluted. Later in the attack, the yolk of an egg may be added. Very feeble patients may have beef-tea with claret, or perhaps Leube's pancreatized beef enemas.

Stumpell recommends milk, with coffee, brandy, or cocoa added: Nestle's milk food; broth thickened with rice or sago; zwieback: with an egg or raw beef for very feeble cases. Beef tea is strongly recommended. Meat peptones may be sometimes useful.

None of the writers quoted appear to have a clear conception of the pathological conditions present, or of the physiology of digestion. Soups depend on gelatine; milk on casein; the yolk of eggs on albumen and fat; the carbo-hydrates mentioned on unconverted starch; beef tea on innutritious extractives. None of these are suited to the conditions present. Liebermeister is right in insisting upon the liberal use of water. The digestion is thereby improved, and the emaciation largely prevented. If patients do not ask for it, water should be given systematically, in stated quantities. Milk should be administered only when predigested. Recently I have employed the Fairchild peptonizing powder, and with good results; better than when employed for children's diet. Kumyss is better still, when some stimulant is required. The raw white of egg, treated with pepsin, and dissolved in ice water, is always acceptable. As this is wholly digested in and absorbed from the stomach, it is especially applicable. Junket, milk digested with rennet, comes under the same category. To these I will add one of the manufactured foods,— Bovinine. I do not wish to be understood as condemning other foods of this description, but simply to state that my experience has been largely confined to this one. I began its use years ago, and the results have been so satisfactory that I have not found it advisable to experiment with others.

Notwithstanding the patient's condition, he will often note the nature and taste of the food offered ; and for this reason, as well as others, it is best to vary the articles given. It is my custom to alternate the use of Carnrick's food, peptonized white of egg, and junket, giving a cupful of either every two hours, in alternation, and with each one a small quantity of Bovinine : from ten drops to a tablespoonful, generally a teaspoonful. This may be given in the other food, or in a little porter sangaree. Another most useful article is coffee, made with milk instead of water. With this a dose of pepsine must be given, say a teaspoonful of Procter's wine. The Carnrick's food consists of milk already digested, and of soluble carbo-hydrates, so that all its elements can be absorbed speedily from the stomach. The Bovinine consists of beef's blood, egg-albumen, whiskey and glycerine.

The action of this food is peculiar and somewhat complicated. As a food, nothing is so readily absorbed as egg-albumen and blood, as nothing comes so near the composition of the human blood. The glycerine assists in keeping the bowels soluble; the whiskey is a useful stimulant, and the boric acid assists the antiseptic remedies with which most practitioners now treat typhoid fever. But there is something more than this in Bovinine. Some years ago I mentioned this and cal ed attention to the fact that in blood we have a substance that has been not only digested and assimilated, but *vitalized*. It is a living fluid, whose existence is identical with that of the individual in whose arteries it flows. I speak simply as a clinical observer; but I feel sure that when the science of biologic therapeutics has progressed a little farther, we will be furnished the reasons for my present claim, that there is in blood as a food, a value not wholly explicable by its chemical composition. Stern has shown that human blood-serum is destructive to the Klebs-Eberth bacil'us, and that the serum of persons convalescing from typhoid fever has an attenuating effect upon the toxicity of typhoid bacillus cultures. The effect of the serum of animals insusceptible to typhoid fever seems to be the next step for investigation.

The net results of the application of the diet herein recommended are these: 1. Avoidance of the gastro-intestinal irritation due to undigested food. 2. The sustaining of the patient's strength, by really feeding him (as distinguished from the mere placing food in his stomach), and the consequent avoidance of collapse, and all the long train of ills that come from mal-nutrition. 3. The avoidance of the excessive emaciation so often seen after protracted attacks of typhoid fever. 4. Shortening of the convalescent period. 5. I put forward, tentatively, my impression that the secondary degenerative lesions of muscles, nerves, and other tissues, are not wholly due to continued high temperature, but, in part at least, to innutrition; and that these lesions are not nearly so marked when the patient has been fed upon the system herein advocated.

TWO ILLUSTRATIVE CASES WITH PECULIAR SYMPTOMS.

In the latter part of August, a young man employed at Cramp's ship-yard applied to me for relief from a severe pain in the head, following a heavy blow on the left temple. The pain was most severe at the site of injury, and close examination detected a slight depression of the frontal bone at that spot. The pain did not sub side, although he was confined to his room and kept on low diet, with cold to the head and arterial sedatives. In a few days fever manifested itself, worse at night, with a slight tendency to delirium, together with tympanites, some diarrhœa, tenderness in the abdomen, anorexia, and a typical typhoid tongue: "small, pointed, red at the tip and edges," with a tendency to dryness, and brownish along the middle. The typical typhoid spots appeared a few days later, and the urine responded to Ehrlich's test. The onset of the specific fever had been masked by the symptoms resulting from the very severe injury received on his temple. This accounts for his not being placed on the sulpho-carbolate of zinc at the beginning of the attack. About the fifth day of the attack he was placed on this treatment—his temperature at that time being nearly 105° in the afternoon. He was given one of Upjohn's 2½ grain, keratin coated pills every two hours, until the stools became natural; after this he took the pills every four hours until the temperature returned to normal. The usual results followed the administration of the zinc salts; the temperature fell to 103°, and never afterwards exceeded that height, ranging from 101° to 102.5°, as a rule. The abdominal symptoms subsided, and were not marked during the rest of his illness. The headache became less after the first week, but did not entirely leave him until convalescence was fully established. Successive crops of rose spots appeared.

The patient was under treatment for thirty-two days; so that the duration of the case was not materially shortened, although its severity was much ameliorated. During the first three weeks he took scarcely any nourishment excepting Bovinine, of which he was given a teaspoonful every two hours, night and day.

There was not nearly as great a degree of emaciation as is usual after attacks of typhoid fever, lasting nearly five weeks.

On November 17th, six weeks after he was discharged from treatment, he reported, looking and feeling very well.

As supplementary to the foregoing article, we add the following remarks by the same authority from the MEDICAL TIMES AND REGISTER, Jan. 6, 1894.

I have several times called attention in this journal to one of the most important problems presented to the practitioner—the feeding in typhoid fever. In Europe the practice has reverted generally to the Hippocratic diet of water-soup.

In America milk is all but universally used.

Milk, according to Dujardin-Beaumetz, can only act as salt and water, as neither the fat nor the casein can be absorbed.

The disease affecting the glandular apparatus of the intestines, absorption through this channel is impossible, and the patient can only be nourished by means of absorption through the veins.

That this is insufficient is shown by the cases occasionally seen of occlusion of the thoracic duct, in which the patient becomes greatly emaciated.

In fact, this condition is exactly paralleled in typhoid fever, where the glands drained by the thoracic duct are rendered incapable of absorbing food. The only exceptions to this rule lie in the facts that all of Peyer's glands may not be wholly disabled at the same time, as the glandular affection is somewhat progressive from above downwards, and some of the glands may not be affected at all.

It becomes, then, a question whether we can supply food at all during a typhoid attack; whether any substance can be directly absorbed into the veins without passing through the intestinal glands and yet be assimilated.

There are two substances in which this may be possible. Egg albumen is directly absorbed into the tissues of the growing chick, without digestion or assimilation. The food is the life; the digested, assimilated and vitalized final product of the whole chain of processes by which food becomes transformed into an integral part of a living organism.

If any substances are available in these cases it must be these. Even milk must be digested before it goes to nourish the child.

Several years ago I presented this subject, and spoke of the excellent results I had obtained from the use of these foods in typhoid fever.

The white of egg can be mixed with iced water and given very readily.

For blood we must rely on Bovinine, as fresh blood cannot possibly be obtained at the times it is required.

Bovinine, consisting of beeves' blood and egg albumen, preserved with glycerine and whiskey, with a little boric acid, answers the need most admirably. It has been my reliance in feeding to typhoid cases for many years, and its success has demonstrated the correctness of the above propositions.

Fourteen drops to a teaspoonful may be given every two hours, day and night.

Patients fed on Bovinine get up with much less emaciation than those fed on soups or undigested milk.

Quite recently a very remarkable series of cases have been reported, in which chronic ulcers, even of many years' duration, have been cured by the local application of Bovinine. Several hundreds of such cases have been so treated with great success. These go so far to confirm my views ; for if Bovinine can be absorbed from the surface of an ulcer, or from the subcutaneous tissue about it, and so improve the local nutrition as to bring about healing, how much more likely that such a substance can be absorbed from the stomach, and keep up the general nutrition.

I would like to know the experience of others in this matter. Too often the good results one gets, or thinks he gets, are not confirmed by the common experience of the profession, and thus error finds credence. It is not what one person, specially skilled in the use of a remedy, can do with it, but what the average doctor who has no special experience with, or liking for the remedy, can accomplish with it ; that is the true test of its utility, and this I wish to obtain.

834 Opera House Block, Chicago, Ill.

SOME CASES OF EXTREME EXHAUSTION.

South Easton, Pa., Nov. 15th, 1887.

GENTLEMEN:—I have recently used your Bovinine in two typical cases which I will now relate. The first was a case of prostration during severe diphtheritis, where it sustained and restored strength in such measure as to astonish the friends as well as myself. The child, after being deemed dead by the friends and neighbors for nearly half an hour (so they told me), rallied, and when I called in the morning was sitting up. In five days she ran about the house.

The second bottle I gave to a case seven miles out from the office, on account of great debility, four weeks after a confinement, owing to the patient getting around too soon. I could find no disease, except debility, and waited patiently for a summons to go seven miles against a north wind. None came, and when I met the husband three weeks later he assured me the bottle of Bovinine did wonders for his wife. I fully believe that Bovinine caused the cure in both these cases. Respectfully,

A. H. R. GUILEY. M. D., 413 Centre St.

Cincinnati, Ohio, Jan., 1888.

GENTLEMEN:—In the case of a girl 14 years of age, from whom a large intraligamentous cyst, with extensive adhesions to the liver and stomach, was removed: vomiting commenced on the third day while the patient was taking milk and lime water alternately with whiskey and hot water. The stomach was rested for 12 hours, after which Bovinine, one dram in half an ounce of water, was substituted for the milk and lime water. The vomiting ceased and did not recur. The digestion of Bovinine is not attended by the formation of gas, as is observed to result from the digestion of milk. This is a consideration of importance in conditions where tympanitis is a troublesome symptom. CHAS. L. BONIFIELD, M. D.

Brooklyn, N. Y., Feb. 14th, 1888.

GENTLEMEN:—Some three months since, while I was slowly recovering from a severe attack of diphtheria, my attention was called to your fluid food, Bovinine. I had no faith in such preparations, but was determined to find out in my own case if the high praise given to it was well founded. I took a large tablespoonful four times a day diluted and flavored to suit for nearly three weeks, by which time my anæmic condition had almost entirely disappeared, and at the same time a careful microscopic examination of my own blood both at the beginning and at the conclusion of my use of Bovinine showed an increase of nearly 35 per cent. of red blood corpuscles in this comparatively short time. Bovinine discloses large quantities of blood corpuscles floating in its albuminous fluid by the use of a first-class microscope with one-quarter inch objective and a B eye piece.

HENRY TUTHILL HALLECK, M. D.

61 Fleet Street.
St. John, N. B., Mar. 31st, 1887.

GENTLEMEN:—I take great pleasure in testifying to the beneficial results consequent upon the extended use of your Bovinine. Having used it in a variety of cases, I have no hesitancy in recommending it as an excellent nutrient, and I feel confident that its use will largely increase among practitioners who are sufficiently unprejudiced to recognize its superior merits.

I have proved its efficacy in the following cases: uterine hæmorrhages, exhausting diarrhœas, nervous prostration, excessive irritability of the digestive organs, hæmorrhages of the lungs, obstetrical cases, convalescence from fevers and the wasting diseases of infancy, such as anæmia, marasmus, cholera infantum, inanition and infantile atrophy. It is especially beneficial in diseases of women where there are exhaustive demands on the system.

W. S. MORRISON, M. D.

DISEASES OF CHILDREN

BY

JAMES C. OAKSHETTE, M. D., C. M.

Post-Graduate Medical School and Hospital, Chicago

—

Coincident with the beginning of the new life commences the struggle for existence. Battle is immediately waged by innumerable foes, and is maintained until the close of that life.

Among the many ills peculiar to infant life may be mentioned, first of all, the inherent, nerve-tiring diathesis, or tendency to certain forms of degeneration, constantly sapping the fountain springs of healthful life, and handicapping it from the very commencement. A weakness due to certain forms of disease, as for instance, Syphilitic, Scrofulous or Tuberculous conditions, so commonly met in daily life : how far-reaching and all-encompassing are these diatheses, the veriest tyro in clinical work will fully appreciate.

Resultant from the diathetic tendency (in many cases) we find the child falling an easy prey to various forms of acute troubles, among which are Cholera-Infantum, Stomatitis, Chicken-pox, Measles, Scarlet Fever and Whooping-cough, or the more deep-seated, and therefore more formidable chronic complaints, of which, prominently before us, are Marasmus, Rachitis, Inanition, Chorea, Glandular affections and Blood diseases: all of these are, alas, but too often aided and abetted in their work of destruction by the gross ignorance of even the simplest safeguards on the part of those who, in the natural order of things, should stand for the guardians and protectors of the infant life committed to their care.

Over against this formidable array, most frequently single-handed, stands the physician, of whom, in the light of clinical experience, it is quite within the truth to assert (despite the many failures) that he is more than conqueror ; and if anything can be offered that will assist him in his warfare, add to his armamentarium or enhance his power, thereby giving him the more easy, sure and unqualified supremacy, our purpose will be served. Leaving for abler men to discourse on the more complex substances and potent remedies, we shall confine our attention to the question of fortifying the citadel against attack, by means of ample nutrition.

Nature has furnished us with a typical, ideal food for the sustenance of its young, and has supplied the mother with a pabulum suitable for the preservation of her offspring, in the form of milk. The study in milk, then, should be, and must be, our standard in dietetics especially prepared for the nourishment of young children, and the nearer we arrive at its chemical and physical properties, the more adaptable will be our food for the nourishment of infant life.

Prof. Tarnier describes milk as a substance composed of two parts,—a liquid portion derived from the serum of the blood by transudation, and a solid part suspended in the liquid, consisting essentially of fat globules. The liquid part of milk is mainly water, containing various substances, such as sugar of milk, nitrogenized matter, inorganic salts, gas, etc., in solution.

The proportion and composition of milk varies very greatly among the several orders of the Mammalia, no two ever producing milk of the same chemical composition ; it also differs widely in this respect among different individuals of the same species, this being brought about by the conditions under which they live. But while the proportions apparently differ in the various species, yet the composition is always constant between well defined lines, and by taking the average composition of a large number of samples, we can arrive at definite conclusions. The essential constituents of milk are Casein-Albumen, Fat Globules and Sugar of Milk, and amount to about thirteen per cent. of its weight. The Casein-Albumen is by far the most important factor of all, but these three substances with a small proportion of inorganic salts and water, are absolutely necessary for the sustenance of healthful life.

The elaborate experiments of Voit, Ranke and Munk have proven, beyond a doubt, that the Albumenoids, or Proteids, are absolutely essential to the welfare of all flesh-eating animals, and that while they are able to exist for an uncertain period upon farinaceous substances, yet sooner or later the inevitable result will come, and death from semi-chronic starvation must, of necessity, ensue, owing to a lack of sufficient albumenoid matter for the maintenance and regeneration of the protoplasm and secreting cells.

From the foregoing statement it will be evident that food prepared from farinaceous substances can never be really applicable for proper nutrition of infant life, inasmuch as it does not conform to the standard, viz., the composition of milk; for when starchy matters are acted upon by "diastase" they are converted into Maltose and Dextrine, not into Albumenoids and Nitrogenous substances such as the standard calls for ; and while it cannot be denied that children do live and thrive on these artificial foods, yet our attention must be directed to other channels before we can produce a suitable pabulum, perfectly adapted to the nourishment of the infant life, and until we can produce some such food more nearly approaching the composition of this lacteal fluid, we must admit the too evident fact that our food is faulty in construction, and hence productive of disease.

In your daily practice you will be brought face to face with this condition of affairs. Cases will be presented demanding definite, decisive and speedy action, leaving no room for an experiment, no time for delay. *We must act.*

The child has been sick for days, perhaps for weeks. Its strength is exhausted, its vital resources drained, its little hold on life is lessening with most alarming rapidity. Its parents have called you to save it. Tears are falling; the home is threatened; their hearts are breaking; they read your face with steadfast, searching gaze. May they hope? Doctor, what will you do? Let the experience not of the one, but of the many, guide and counsel.

Dr. Caldwell Smith (in the Hospital Gazette) says: "Nitrogen, which is one of the main constituents of Albumenous or Nitrogenous foods, is present in the brain, the nerves, the blood, and some must therefore be supplied to make up for the waste which goes on constantly in the system." * * * "If too little of this form of nourishment is supplied, loss of weight, bloodlessness, want of energy and life result." The very conditions of your patient, Doctor.

Because of its wealth of Nitrogenous elements, eminent colleagues all over the world, in all climates, in all conditions, have given in such cases, "Bovinine." It meets the indications, saves the patient, and establishes your reputation.

In my own experience, this article "Bovinine," stands pre-eminent above all other foods for these cases, and not these only, but all where the largest amount of nutrition in the smallest bulk, the least irritating and at the same time the pleasantest form, is the desideratum.

Bearing in mind the above statements, and in the face of the experience of many, it is at once a pleasure and a duty to direct attention to this well tried and thoroughly nutritive agent, representing more nearly the ideal nutritive force than any other within our reach to-day. Among its commendable features may be mentioned its rapid absorption and ready assimilation, approximating the transfusion of blood into the system, rendering a happy response to its administration almost immediate, at once gratifying to the physician and parents; it is thus restoratively healthful and sustaining to the patient, even though he be bordering upon the confines of another world.

81

PHYSICIANS' CASES IN CHOLERA INFANTUM, MARASMUS, ETC.

Indianapolis, Dec. 16th, 1887.

GENTLEMEN:—During last summer, while attending a case of cholera infantum in a child of six months, I used your Bovinine with most satisfactory results. After exhausting all other methods of nourishment to no purpose (they being either rejected or passing undigested, and my patient being at point of death from starvation), I happened to think of a bottle of Bovinine left with me a short time before. Without much hope of benefit I commenced the use of it in small doses diluted with the mother's milk. To my surprise it was retained and perfectly digested. At once a great change was apparent in the little sufferer. I continued the Bovinine in larger doses and with the most gratifying results. The vomiting ceased, the stools were more natural, showing no undigested food in the passages—in short, I attribute to your Bovinine more than all other treatment the cure in this very critical case.

S. H. MOORE, M. D., 152 Virginia Ave.

Philadelphia, Pa., Oct. 2nd, 1887.

GENTLEMEN:—About July last, during the excessive hot weather, I was called to see a child three months old, that was rapidly wasting away from that dreadful disease, Marasmus. My attention had previously been called to Bovinine, and I at once ordered the same to be given in doses of five drops with a tablespoonful of water every hour. Improvement was almost instantaneous, and the dose was gradually increased to fifteen drops, and this treatment kept up for four days until a wet-nurse was obtained. In about a week's time I had the pleasure of seeing my little patient bright and sprightly, who but a week previous had been given up by another physician to die. Very respectfully yours,

W. W. JAMES, M. D.

Wilmington, Del., Oct. 12th, 1887.

GENTLEMEN:—I cannot too sincerely express my thanks for your Raw Food Extract, Bovinine, to which my attention was most opportunely and fortunately called. My babe was sick all Summer of cholera infantum, which had emaciated and so wasted it as to produce marasmus, and it was, when first taking the food, in an almost helpless and hopeless condition.* It is now thriving and growing, being amply nourished, and a living triumph for proper food intelligently administered. I wish you the success which your preparation so richly deserves. Most respectfully yours,

J. R. C. GORRELL, M. D.

Pittsburgh, Pa., Oct. 7th, 1887.

GENTLEMEN:—My attention was first called to Bovinine in the latter part of August of this year. From that time I have given it in quite a number of cases of cholera infantum and the so-called summer complaint of children, and have been gratified beyond my expectations with the results. I can recall several cases where impending death from inanition was averted beyond all question by the Bovinine. No other food that I am acquainted with (and I think I have tried nearly all) has given me the uniform good results that this one has. W. J. MARTIN, M. D.

Richmond, Va., Nov. 15th, 1887.

GENTLEMEN.—I have recently used your Bovinine with the most gratifying results. A child who had suffered for many days with the most distressing irritability of stomach, and to whom a number of tried remedies had afforded no relief, after the first dose of Bovinine quietly slept, and waking with no return of nausea, made rapid recovery. Thus, I am prepared to say that, in addition to its superior food qualities, so easily assimilated, Bovinine possesses the great merit of being acceptable to the palate, and is tolerated by the most irritable stomach. Truly yours,

SAML. D. DEWEY, M. D.

Allegheny, Pa., Jan. 12th, 1888.

GENTLEMEN:—I have used your Bovinine and found it to be everything that is claimed for it, finding it especially useful in fevers or other diseases where a fluid diet is necessary. It seems to have stimulative as well as nutritive powers, and is very readily assimilated. In very exhaustive diseases of all kinds it has wonderful restorative power and deserves the highest indorsement the medical profession can give it. Respectfully yours,

G. A. MUELLER, M. D.

Wilmington, Del., Sept. 26th, 1887.

GENTLEMEN:—We have found Bovinine more than satisfactory in a number of cases. In cases of mal-nutrition, prostration and exhaustion, it is the remedy par excellence. It is without doubt the most efficient of all raw food extracts.

L. & L. A. KITTINGER, M. D., 724 King St.

FEEDING IN SOME FORMS OF MALNUTRITION.

BY

WILLIAM F. WAUGH, A. M. M. D.,

PROFESSOR OF CLINICAL MEDICINE, CHICAGO
POST-GRADUATE COLLEGE.

To all such remedies as iron, manganese, the phosphites, oils, etc., the reconstructives, in fact, the objection has been urged : that these substances exist in the common, every-day food, and that, in quantities beyond the needs of the system, as a part of what is taken may be found in the fæces. If this be so, why give more, when the body cannot utilize what is already taken? But this is only a bit of *a priori* reasoning, of which clinical observation has long since demonstrated the futility. For we may give a grain dose of iron and find the stools blackened by it, showing that the larger portion is not absorbed ; nevertheless, if we give ten or thirty grains, we will find vastly more of it is absorbed, and the chlorosis is much more rapidly cured than by the small dose.

The same is true as to the administration of manganese and cod-liver oil. In the latter case, the presence of fatty acids renders this the most digestible of fats, so that it cannot be replaced by cream, butter or any ordinary fat.

Another principle enters into consideration when we take up the phosphites. It is questioned if any of the ordinary phosphorus combinations can be assimilated at all by the human body, and claims have been put forward for the hypophosphites as being peculiarly fitted for appropriation, by reason of the instability of their chemical composition and the readiness with which they are attacked by oxygen. A similar plea has been put forward for the lacto-phosphates, as they are so freely soluble in water that the chances of absorption are very great. But others express a preference for powdered bone dust, claiming that we have in it the phosphorus-lime combination as it exists in the human body, and that even in its finished state, digested, assimilated and vitalized.

In all this statement we see the lines of reasoning finally converging at the point of *use;* the ultimate, crucial test.

The world is no longer satisfied that a person is fed when food has been put into his stomach ; it is necessary to follow that food to its ultimate destination and see if it be digested, absorbed, assimilated and finally become a constituent part of the human organism, or if it be arrested at any of these milestones and wander off into the byways of pathology.

In affections whose pathological processes are carried on notably in the alimentary canal, these considerations are of special importance. Mothers must be taught that their infants will die of choleraic adynamia, no matter how well filled the stomach may be with wholesome food, and the evacuations penned in by opiates. Great as is the necessity for food to sustain the strength during these exhausting attacks, it would be better to leave the child without food than feed it as many do, on raw milk not always beyond suspicion, bread and sugared beverages.

When the intestinal mucosa are in a state of acute catarrhal inflammation, the ordinary processes of digestion are interrupted, and food ordinarily bland becomes intensely irritating to the inflamed viscera, especially if the food undergo decomposition. Hence the indication is for food that in the smallest bulk contains the largest amount of nutriment, in a condition that allows of its speedy absorption. For if the food is likely to be soon ejected, it must be predigested, that some of it may be absorbed before vomiting occurs. Nothing answers so well as the raw white of egg ; for this is already prepared for absorption into the tissues of the embryo, without passing through any digestive or glandular systems whatever ; mixed with ice-water, it is as unirritating as any food can be and if it only remains a few minutes in the stomach a part will in that time be absorbed.

In typhoid fever we see also the digestive processes inhibited, but the stomach usually allows food to be retained. I 'am always sorry to see an otherwise sensible article on the management of this disease close with the recommendation of "milk and light farinaceous food." The casein cannot be digested, nor the fat emulsified, nor the sugar converted, as there is no secretion of the peptic ferments, nor could the fat be absorbed, as the disease of Peyer's glands blocks up the lacteals as effectually as if the thoracic duct were occluded. If the glandular affections be general, milk can only act as salt and water ; and the remaining constituents, incapable of digestion and absorption, remain in the alimentary canal to decompose, and thus increase the patient's danger and distress.

The proper foods in typhoid fever are those that can be absorbed directly into the stomach-veins without requiring digestion, as below this viscus little or no digestion is performed.

Tne nearer we come to blood itself, the more likely we are to nourish our patient. The raw white of egg is of use as far as it goes, but in an attack lasting as long as a typhoid some more general diet is required   Nothing would answer so well as pure blood if we could get it. We have it, however, in Bovinine, in a form that amply fulfills every indication—bullocks' blood and raw white of egg, preserved in glycerine and whiskey, with a little boric acid to prevent decomposition after the bottle has been opened. Here is all we want, without any objectionable ingredient. Even Davis could scarcely object to the amount of whiskey present in a tea-spoonful (about 12 drops), while most physicians would be more likely to reinforce this dose. In all my cases of typhoid fever treated during the last few years, I have relied on Bovinine as the principal food, giving up to a tablespoonful every two hours. There has been very little emaciation in these cases, far less than in those treated without Bovinine. In the first week, and after con-valescence has begun, I give junket, or the best predigested foods in the market, and no other nutriment is given or required.

The advantages of this diet are : the task of alimentation is reduced to a minimum ; the smallest possible quantity of food is to be swallowed ; the food is pure nutriment with no waste ; the food is such as to require absolutely nothing of the patient except absorption ; the nutritive value of the diet is so high that the patient is not compelled to live on his own tissues, and emaciation with debility during a prolonged convalescence is avoided.

Not the least of the advantages of Bovinine is its ability to keep well, without decomposition, in any sort of weather with any sort of care from nurse, parent or friend, without special education or intelligence. For a system may be almost ideally perfect, and yet of no practical value because its carrying out requires a degree of perfection in the nurse, never to be secured outside of the professional ranks.

I am aware that very good results are obtained with the milk diet, but my remarks are intended for those who are never satisfied with a good result if they can obtain a better.

In the use of Bovinine in cholera infantum, marasmus and other diseases of children in which malnutrition plays a leading part, I am always reminded of the soldier who hung a horseshoe about his neck on going to battle. When a bullet flattened itself against the horseshoe he remarked, "But little armor is necessary if it be in the right place." Very often the vital power and the force of disease are so nearly balanced that there is but a narrow margin between life and death. But the effect of five-drop doses of Bovinine every half hour will overcome an adverse balance larger than any one would believe possible who has not watched its effects.

Give a little aid to the bodily powers, invigorate the leucocytes with a steady supply of reinforcing albumen and hæmoglobin, and the winning rally becomes possible. With richer blood the digestive glands elaborate more of their products, better digestion of food becomes possible, and thus the effect is confirmed.

There is a nicety about such feeding that commends itself to one who believes in accuracy in medication, in precision wherever possible, rather than in the slovenly practice of "general principles" and nothing more.

SOME FURTHER AUTHORITIES.

Dr. Burwell's careful investigations and experiments with the various foods for infants give great weight to his conclusions. The moral to be derived from his article is that Bovinine is always valuable, having within itself life-giving properties which may be lacking in any other food. Therefore, to insure full and perfect nutrition and avoid the deadly Summer diseases of infants, use Bovinine in connection with any food you may be feeding to your child.

In "The Medical Bulletin," Philadelphia, the Doctor says :—

" I have found the most satisfactory food for children to be cow's milk sweetened with sugar of milk, diluted from one-third to one-half, to suit the age and condition of the child. I generally add a peptonizing tablet and a few drops of Bovinine.

"The addition of Bovinine makes the mixture more nutritious, for Bovinine is more readily digested than any other food substance and supplies the cow's milk with a pabulum at once highly nutritive and easily digested."

He also gives the following report regarding the use of Bovinine in the Washington Foundling Asylum :—

" During my attendance at the Washington Foundling Asylum I had occasion to use a mixture of cow's milk and Bovinine in a large number of cases and met with uniform success. In several instances of extreme emaciation, and where the child was unable to assimilate the slightest quantity of cow's milk, I gave the Bovinine alone by *putting a few drops on the tongue.* It was surprising how readily the little sufferers responded to this treatment ; they at once brightened up and soon were able to digest the usual cow's milk and Bovinine mixture. Several marked cases of marasmus were treated with Bovinine, milk and brandy, with an inunction of cocoa butter and alcohol. In two cases at least, where the burial clothes were kept in readiness by the head nurse, the children, to my surprise, rapidly gained strength under this treatment, and a year later one of them (little Annie Bennett) was adopted from this institution a plump and healthy child."

Dr. Herman C. Marcus, in the "Times and Register," another medical journal in Philadelphia, says in the treatment of invalids:— "Whatever food be given, the quantity should not be large ; say a small teacupfull given every two, three or four hours. A large bowl of broth will sometimes bring on a violent headache. If the amount of nutriment in a cupfull be insufficient, it is best to increase the nutritive value without increasing the bulk. This may readily be done by adding Bovinine, from a few drops to a teaspoonfull to each cup of food." He mentions several cases to illustrate its value in this respect. We print one for example :—

"A child eight months old had ceased to nurse and all efforts at feeding occasioned such pain that the struggle still further exhausted the child Bovinine was ordered in ten-drop doses every half hour, and as the baby took it readily the dose was increased to half a teaspoonful. This was the only food taken for two days, the child refusing to take the blandest food known, the white of egg. It is doubtful if the child would have borne up under its afflictions had it not been for Bovinine. These cases were mentioned as types of a large class in which the addition of Bovinine to the other means employed was the one thing needful to tide the patient over the critical period."

The "Times and Register" also says Bovinine should not be confounded with other preparations, some of which are totally unfit for use in a sick room. The article concludes by remarking : —"It is sufficient to say that no other liquid food preparation has been tendered the medical profession of equal nutritive value with Bovinine, whose composition and method of preparation are made public, so that physicians in employing it may know exactly what they are giving. Until some other preparation approaches the fulfillment of these conditions, Bovinine cannot be said to have any rivals."

Dr. Edward P. Vollum, U. S. Army Medical Director, reports a large number of cases of gastric irritation, dyspepsia, nausea and distressing sensations of fullness and weight in the stomach after eating. All of these cases were promptly relieved by Bovinine taken soon after eating. He says :—"I was greatly surprised at the anodyne effect of Bovinine on the stomach when I first noticed it, but the explanation of the action would seem to be that the Bovinine contains the elements that the stomach in its moments of distress needs."

Dr. J. Wesley Bovee, of Washington, reports a number of instances in which Bovinine was given with marked benefit at the Washington Asylum Hospital in cases of wasting diseases, and of aged people who seemed to pick up and get vigor from its use.

Dr. J. R. Wellington, of the Children's Hospital at Washington, permits our reference to that institution as having used Bovinine with excellent results. He mentions one interesting instance of "Gastrotomy" where the child was fed with Bovinine for days *through a tube inserted through a hole in the stomach.*

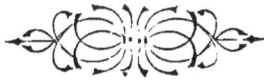

CHLOROSIS.

BY

DR. HERMAN D. MARCUS,

Late Resident Physician at the Philadelphia Hospital.

Chlorosis is a disease of the blood resulting from a diminution of red blood corpuscles and hæmoglobin. Examination shows a decrease of nearly half the red corpuscles, which takes place slowly, while the hæmaglobin becomes rapidly decreased to about 25 or 30 per cent. of the normal amount.

Chlorosis is generally a disease peculiar to the female sex, at the age of puberty, and may be predisposed by congenital and functional disorders of the circulatory and sexual apparatus, which at times furnish accompanying symptoms of chlorosis. Organic diseases only rarely cause this condition.

One of the most prominent symptoms noticeable is anæmia. The skin becomes of a peculiar pallid, greenish color. The decreased oxygenation of the blood causes more or less depression of all the organs, and the patient becomes easily tired under slight exertion, complains very much of malaise and lassitude, is continually drowsy. The mental activity decreases owing to the mal-nutrition of the brain. The patient sleeps considerably and a certain amount of mental torpor is noticeable. Very often excruciating headache is complained of. This in itself may be the symptom which brings the patient to the physician and may manifest itself in migraine or as a dull pain felt in any part or over the entire head. Other symptoms such as muscæ volitantes or tinnitus aurium may be also observed. The appetite of the patient is very much decreased and it is at times necessary to use extraordinary means to prevent inanition. The bowels are generally constipated.

The circulatory system shows a number of symptoms such as the loud blowing sound noticeable in the veins of the neck, the so-called *bruit de diable*.

The excitability of the vaso-motor system is also peculiar to chlorosis, causing blushing under very slight provocation. It would be well to remark here that chlorosis is not of necessity accompanied by pallor. The other organs such as the liver, kidney and spleen, show nothing abnormal. The urine shows very little difference from its normal status.

Menstruation may be abnormal, and if so, it is due to the changes in the blood vessels.

Chlorosis, as mentioned above, affects mostly the female. Men may appear chlorotic, but when such is the case we are generally able to discover a tuberculosis of the apices crurum, medulla oblongata, abdominal organs, or some nervous affection due to overwork.

Constipation, which is so commonly a symptom, is frequently the underlying cause of chlorosis. This may be explained by the decomposition of the fæces causing an absorption of ptomaines into the circulation. This theory receives strong support by the fact that a number of cases may be cured by thorough purgation, without the use of such therapeutic agents as arsenic, iron, etc. This explanation may be rather far-fetched, but recognizing the possibility of such a theory we can in the absence of other explanations do nothing better than accept its probability.

The prognosis of this disease is in the majority of cases favorable. But it must be remembered that quite a number of complications may arise during its course. Nervous symptoms may manifest themselves, such as migraine, neuralgia or pains in various parts of the body, causing a lessening of the power of resistance of the nervous system. This may result in time in neurasthenia or at least hysteria ; such a result is of course very unfortunate, as long-continued and persevering treatment will then be necessary to save our patient a lifelong suffering.

Regarding the therapy of chlorosis we must in the first place pay strict attention to the action of the bowels;—this becomes important whether the constipation is the cause or only an accompanying symptom. Regularity of the action of the bowels is all important and any purgative may be employed, such as rhubarb, aloin, podophyllin or any mineral water having laxative properties. A very good prescription is: —

> Extr. Stilling fl ⎫
> Tr.   Belladonna ⎬
> Tr.   Nucis Vom. ⎪
> Tr.   Physostig. āā, ℥ ii. ⎭

M. Sig. Twenty drops in water before meals.

Above prescription is highly recommended by Prof. Bartholow of this city for habitual constipation.

Next to the condition of the bowels, the appetite demands attention. Nearly all chlorotic patients suffer from anorexia, and this condition must be remedied. Iron is here our best appetizer. This agent has however found a great many opponents to its use, they claiming that the small amount of iron which is necessary to augment the amount of iron in the red corpuscles is taken daily with the nourishment.

The amount of iron present in meat, eggs, etc., is sufficient to give chlorotics enough iron. It is therefore unnecessary to introduce iron into the system as a medicine. Still the fact remains that with the use of iron, the symptoms of chlorosis disappear quicker than without it. We notice under the influence of such treatment a reaction of all the functions. Very soon hæmoglobin is increased as well as the red corpuscles. Hydrochloric acid which was diminished in quantity is again present in its normal amount and causes thereby an improved digestion, and with the return of the appetite, absence of the prevalent dyspepsia; the other functions which were very much disturbed during the progress of the disease, become normal and the patient soon regains her health.

The number of iron preparations are so large that it may become a question which form to employ. We find in the pharmacopia such preparations as Citrate, Carbonate and Sulphate and Iodide of Iron (syrup) etc. The Iodide of Iron is not as efficient owing to the small amount of iron present. The preparation to be employed depends largely upon our patient. Very many are unable to take it in the form of pills or powders, and in such cases we must prescribe tinctures. Another objection is its action on the teeth and mouth, which is very objectionable to the female patient. Pills or powders will be found a very convenient form of prescribing it, but where the liquid preparations are preferred, it would be well to caution our patient to use a glass tube. The dose should be rarely more than 6-8 grains daily, owing to the length of time during which it must be administered. This dose of course means the equivalent of iron in its powder form. It should be administered after meals and continued for a considerable length of time. After 10 to 15 weeks it may again be discontinued to be again taken up if the disease is still present.

Another very important therapeutic agent in the treatment of this disease is arsenic. This drug may be very advantageously combined with iron. Fowler's solution (liq. potass. arsen.) is the preparation commonly used.

The proper nutrition of the patient is one of the most important points in the treatment of chlorosis. Easily digestible food, such as rare meat or soft boiled eggs should be given. It may be necessary to appear most rigid as regards the directing of nourishment, so great becomes the aversion of our patient to food. The appetite becomes at times capricious, in which cases it is well to humor the patient, providing of course that no indigestible articles are craved. It is a very common occurrence in chlorotics to eat sand, chalk and such like, a procedure which of course must not be tolerated. Pickles when asked for may be permitted as they excite the secretion of the gastric fluids.

No substitute for the common nitrogenous articles has been found of so great a benefit as Bovinine. This preparation is a highly nutritious food containing the important elements of defibrinated ox blood, and desiccated egg albumen. It is therefore a powerful factor in sustaining and nourishing the system after it has suffered from debility and disease, and in restoring to the blood its lost hæmoglobin by continually increasing its supply of red blood corpuscles. Whereas meat or eggs may not be well borne by the chlorotic patient, Bovinine when properly given will be assimilated. It is best administered in ascending doses, 20 drops to a dessertspoonful every 3 to 4 hours. It may be combined with milk, wine or whiskey if preferred, but if repugnance is shown to it, a tablespoonful in a pint of warm milk may be given as enema 5 to 6 times daily.

DIABETES.

While it is true that the most learned Ætiologists differ as to the causes of this very distressing disease, it is conceded that the element of nutrition is the *all important factor* in its treatment. Bovinine possesses in a very marked degree all the elements needed in the reparative process of an organism suffering from this disease, being especially rich in the nitrogenous elements that go to replace the waste of the nervous system; it is at the same time the most easy of assimilation and is absolutely devoid of all the elements of starch or sugar. It has been used with signal success in scores of well-marked cases, and with uniformly good results. Tablespoonful doses in milk or Alkaline-Carbonized water 4 to 6 times in 24 hours will prove the above to be correct.

TUBERCULOSIS.

Is Nutrition the Sine Qua Non in Tuberculosis?

Without fear of contradiction, it may be asserted that tissue building (the establishment of healthy cell life) is the foundation treatment of every case of tuberculosis, not only in its incipiency, but in the advanced stages as well.

This proposition is not antagonistic to the germ theory; rather has the knowledge of the existence and important role of the bacillus led us to a better understanding of the necessity of proper nutrition.

Nutrition is the "sine qua non" in all cases of tuberculosis. In some instances its necessity is more evident than in others; in acute cases other needs may be more urgent, but the rule holds good, that the victims of tuberculosis must be nourished. In "thin living and thick dying" we find tuberculosis rampant.

Feeding is not always nutrition. The best diet may not be assimilated—may do harm rather than good. The practice of stuffing, so honestly advocated by some authors not long since, has been rapidly abandoned. Years ago every case of phthisis got a bottle of cod liver oil; now it is given only in selected cases. Nutrients are chosen which can be appropriated, and food is given in such a manner and of such kinds as will favor complete assimilation.

There must be a demand for nutrition before assimilation can be satisfactorily performed. There must be the ability to appropriate food that is taken, else the defective cell in a remote part of the system will profit little thereby. Just here, I believe, is the important point in the treatment of tuberculosis. The best of food and the most reliable nutrients are taken, and still the waste in many cases goes on. There is either a want of assimilation, or a lack of proper food.

There is need for "Respiratory food" as well as for that in the alimentary tract. Oxygen must be taken into the system and the cells empowered to use it in the nutritive changes which we aim to accomplish by proper feeding.—Dr. Wm. Porter,

in the St. Louis Clinique.

The gratifying success attained by the use of Bovinine in cases of incipient phthisis, proves the truth of Dr. Porter's statements regarding nutrition, which are copied above. Bovinine conveys into the system a wealth of oxygen through the medium of the red blood corpuscles existing in such abundance in this raw food, and which are destroyed by heat in cooking in all other beef preparations.

The subjoined remarks are taken from a very exhaustive treatise on "Diet and Hygiene in the Treatment of Consumption," by Dr. Edwin F. Rush of Chicago, a well known authority in such matters.

"The diet of consumptives should consist of animal food of all kinds, fats, oils, milk and cream, butter, eggs, bread, farinaceous preparations, and scientifically prepared raw food extracts. They should be encouraged to eat, and to this end the bill of fare should be as varied as possible, and served in a tempting manner. It is better to eat moderately five or six times daily. The surroundings should be cheerful, the tray and napkin clean, and each article of food daintily served, and small in quantity. A sick person will often eat if a small amount of food is nicely served, where the sight of a large amount would cause disgust.

Beef juices, or raw meat extracts, are very valuable adjuncts in all, and absolutely necessary in most cases of phthisis on account of their immense nutrient force in a concentrated form, the small quantity required, its tolerance by the stomach, and rapid and complete assimilation, whereby the strength and vital powers are quickly nourished and maintained. Periods will occur in every case of phthisis when gastric irritation will preclude the use of ordinary food; then our reliance must be placed in these raw concentrated foods. Many "meat extracts" and "raw food extracts" are upon the market, a large proportion of them containing no food or nutritive properties, for instance, the highly-prized "beef tea," popularly supposed to have great food virtues, but actually containing none whatever, being merely a temporary stimulant, upon which a patient would soon starve. Some of the raw foods are repulsive in taste and odor, or in some manner objectionable. I have used, in my practice, every known raw food extract, but for two years past I have exclusively prescribed a raw meat extract, prepared in this city, and known as Bovinine. This article of food is very rich in all the elements entering into the formation of blood and tissue, is easily borne by the most delicate stomach, of excellent taste and odor, and is rapidly and completely assimilated. I am personally familiar with the mode of preparation of this food, and know that the blood used is from the best of the finest cattle supplied by the great Chicago stock yards. I do not wish to appear invidious, but I prefer Bovinine above all other raw food extracts for its great nutrient qualities, acceptability, and its large percentage of albuminoids (20.56 per cent). I usually administer this food three or four times daily, in doses varying from ten drops to one teaspoonful, diluted with four or eight times the quantity of water, or milk and cream mixed. In conditions of great exhaustion and debility requiring stimulants, the raw food may be added to milk punch or egg-nog. In extreme cases, or in violent stomach irritation, our main reliance must be placed in raw fluid foods, administered by the mouth, a few drops at a time, and in cases where the stomach will tolerate nothing at all, we must then administer the fluid food by the rectum, a teaspoonful of the food to four teaspoonsfuls of sterilized water, injected slowly, thus giving the stomach absolute rest until it recovers its tone.

"Bovinine contains lactic acid, a normal constituent of the muscular tissues. This acid in the presence of pepsine or pancreatine, rapidly digests nitrogenous matter, which accounts for the speedy absorption and assimilation of Bovinine in cases of impaired digestive functions."

Newark, N. J., Dec. 21st, 1885.

GENTLEMEN:—Having used your preparation, Bovinine, I can say my experience has been so satisfactory I desire to recommend it to the medical profession. In my hands, it has yielded unparalleled results in cases of typhoid pneumonia, acute phthisis and gastric catarrh. The patients all express great satisfaction and confidence in this most valuable nutrient.   Respectfully,

E. W. EDWARDS, M. D.. 11 Washington St.

Reading, Pa., Sept. 25th, 1887.

GENTLEMEN:—I have used your preparation, known as Bovinine, in a marked case of phthisis pulmonalis with exceedingly satisfactory results.   Yours truly,

S. L. DREIBELBIS, M. D.

CHRONIC ALCOHOLISM.

In the treatment of *Chronic Alcoholism, Bovinine* has been put to the *severest tests,* and has *not been found wanting.* Begin with 10 to 30 drops in a little carbonized water every 30 to 60 minutes; continue these small doses until the stomach has become settled; then it may be given in milk in from teaspoonful to two table-spoonsful doses, six to ten times in 24 hours. The *continuous* use of *Bovinine* and good milk has cured and will cure the *Alcohol habit.* The Morphine habit can be cured in the same way, for it is a principle well established in the treatment of both chronic alcoholism and the morphine habit that in the ratio that the functions of nutrition can be brought up to a normal standard, in that same ratio will the desire for either alcohol or morphine subside.

Boston, Jan. 7th, 1888.

DEAR SIRS:—It gives me great pleasure to give you information in regard to a case of delirium tremens treated with Bovinine. The circumstances were these: I was called to Mr. C., and found him furiously insane from "Mania a Potu." After the mania subsided, all the acute gastric symptoms were manifested with a total inability to retain food—all but your preparation Bovinine, which I began to use by adding two teaspoonfuls to a goblet of water, of which a teaspoonful was given every ten minutes the first day, gradually increasing the strength day by day, with the most happy results of fully maintaining the patient's strength and preventing the exhaustion which so often sets in, causing a fatal termination of the disease. I am convinced that if physicians would but give a portion of the care and study to your valuable food remedy they do to ordinary medication, they would realize the great value of Bovinine in general practice. Very respectfully,

H. F. BRACKETT, M. D., 61 Indiana Place.

Report on the Use of Bovinine

AT THE

GENERAL HOSPITAL OF MEAUX,

SEINE-ET-MARNE, FRANCE,

By DR. DUFRAIGNE, Surgeon-in-Chief.

[TRANSLATION.]

"Having read the eulogistic testimonials of American and English physicians concerning Bovinine, and this preparation appearing so familiar to the foreign medical profession, we thought it advisable to give it a fair trial, and that, in spite of our general distrust of all the medical specialties coming to our notice from the other side of the ocean."

"Our first trial having been found successful, we felt no hesitation in giving a wider field to our experiments and we must say that these experiments fully confirmed the happy results of our first trial, and now we may safely declare that the medical value of Bovinine, and above all its nourishing value in certain particular cases, cannot be doubtful, for it is an ascertained fact that the Bovinine preparation surpasses in its qualities all the similar so-called nourishing preparations known to us. Now let us relate a few cases which will demonstrate the efficiency of Bovinine beyond all doubt."

"A patient aged 40 entered our hospital in May, to be treated for a diffuse phlegmon of the hand and left forearm. Continued antiseptic irrigation seemed at first sufficient to check the inflammation, but soon the articulation of the elbow and the lower part of the arm became invaded, and two large incisions were made on the lower surface of the forearm allowing that part to be drained, and revealed an extensive destruction of muscular tissues."

"The patient was becoming weaker and weaker, and a large slough over the sacrum with abundant discharge gave further evidence of the considerable loss of vitality. We had then before us a case of septicæmia, and amputation was decided on above the elbow, under very critical circumstances in consequence of the generally debilitated state of the patient, who had not been able to take any food for several days. We had then a good opportunity for testing Bovinine, and to see for ourselves if that preparation could justify its reputation. The patient took Bovinine with ease, and a few days after (having taken nothing but Bovinine) gradually recovered his strength; delirium and fever disappeared, the mental state was better ; the system being less debilitated absorbed more nourishment, and a general improvement followed. The terrible effects of septicæmia were conquered. To-day the patient is able to walk ; he is convalescent."

"This splendid result was obtained by the easy assimilation of such a vital extract as is contained in Bovinine."

"A young patient aged 15, suffering with osteo-myelitis, half of the lower end of the diaphysis of the tibia being involved, we were obliged to remove a large sequestrum of the anterior surface of the bone and to curette the posterior face ; in short to make a deep excavation of that region. The patient being much debilitated before and especially after this severe trial, Bovinine was administered, 15 grammes per day, and for eight days the patient took it well."

"What is remarkable, regeneration of the bones rapidly became apparent, and when the patient could take the ordinary regimen of food we reduced by half the dose of Bovinine. The result is satisfactory."

"Another patient of a delicate constitution, whose left thigh had been amputated, was fed on Bovinine alone for several weeks. By the help of that preparation, the system, which had been so debilitated before the operation, in consequence of a tumor of the knee, which had resisted treatment for ten months, has regained its vigor. The patient will soon be able to leave the hospital in excellent condition."

"Therefore we may condense and say, judging from these cases taken at random, and also from many others which have given the same good results, (and without according to that preparation any therapeutic value), that Bovinine remains a powerful re-building factor for debilitated patients after a long illness; for patients after surgical operations; for patients digesting their food with difficulty; in an accidental or traumatic case, or after a constitutional illness; as the results obtained in our medical service and other wards of our hospitals have thoroughly demonstrated. These results are the consequence of the vital principles contained in Bovinine, principles carefully studied and analyzed by the pharmacists of our establishment with reference to their qualitative and quantitative value, prominent among which are fibrin, casein, hæmoglobin, peptones and the albumen of the egg." (Signed), DUFRAIGNE.

NUTRITION IS THE PHYSICAL BASIS OF LIFE."

"The effects of diet are profound and far-reaching, and cannot be over-estimated. The present indications are that a change in the relations of food and medicine is slowly taking place. The tendency among our best physicians largely to substitute food instead of medicine in the treatment of disease is not to be looked upon as a mere fashion. The dietic movement has a wider and deeper meaning."

SIR WILLIAM ROBERTS,
Address before the British Medical Society.

"CHRONIC STARVATION."

"For the last six or seven years I have tested, by carefully inquiring into the past history of patients mostly suffering from some uterine or ovarian disease, or some affliction incidental to childbed, and these conclusions have stood the test of this long-extended inquiry. I have to state the important conclusion that a continuous insufficiency of food, or what may be called "a chronic starvation," more or less intense in different cases, was found to have existed almost universally. Consequently, I have naturally been led to consider chronic starvation as a most important factor in disease."

"It is generally admitted, by authorities on the subject of diet, that nitrogen is the most essential of all foods, and that a certain amount should be taken daily. It appears that the diminution in quantity of food most frequently affects the nitrogen. Meat is the article of diet which as a rule, is the source of the greater part of the needed amount of nitrogen, for, in England, at all events, meat is the popular article of food, and, in cases of chronic starvation, we mostly find the quantity of meat taken is exceedingly small."

DR. W. M. GRAILY HEWITT,
Address before the British Medical Association.

Bovinine contains precisely the elements indicated by Dr. Hewitt as being highly essential to the proper nourishment of the body, and those in a form most easy of administration and agreeable to the taste.

The Bovinine Company, London, August 22nd, 1802.
GENTLEMEN:—I have the highest possible opinion of Bovinine as an easily assimilable highly nutritious food.

Again he writes:— October 31st, 1890.
I may remark that I have every week additional reasons to speak well of Bovinine. It is one of the few articles which, I believe, will never go out of fashion. —— M. D.

The Bovinine Company, 3 Roundhay Terrace, Leeds,
Oct. 15th, 1890.

GENTLEMEN:—I duly received the sample bottles of Bovinine you were good enough to send, and after taking half one bottle myself I used the rest on poor patients, whose condition called for nutrients exhibited in concentrated form. I am more than satisfied with the preparation—it is one on which the Profession may rely without fear of disappointment. I found its immediate effect to be increased blood-pressure, showing that it is a true nutrient-stimulant needing no digestion, but being at once absorbed. Its invigorating effect in a case of severe choleraic diarrhœa was magical—the symptoms of collapse yielding in a minute or two.

F. ARNOLD LEES, M. R. C. S. Eng. L. R. C. P. Lond.

The Bovinine Company. ——, Ireland, Oct. 27th, 1890.

DEAR SIRS:- I have much pleasure in bearing testimony to the beneficial effects of Bovinine. I used a large bottle in a case of extreme debility, and can safely say it saved my patient's life, when I could not use special stimulants in any other form.

Faithfully yours,
——L. R. C. S. I. &c.

DISTINGUISHED DOCTORS AND PATIENTS.

General U. S. Grant was sustained for months previous to his decease almost wholly by the use of Bovinine, as the following letters will testify.

"During the last four months of his life, the principal food of my father, Gen. Grant, was Bovinine and milk, and it was the use of this incomparable nutrient alone that enabled him to finish the second volume of his personal memoirs." F. D. GRANT,
[Late U. S. Minister to Vienna.]

The Bovinine Company,

GENTLEMEN:—It is a long time since I became acquainted with your excellent fluid food, Bovinine. I have embodied my experience with it in the case of General Grant, in my narrative of "The Last Days of General Grant."

My attention was first called to the preparation by the Hon. Salem H. Wales, of New York, who sent me a quantity which I critically examined before using, and compared with other foods of a similar character which had been most liberally supplied me. The Bovinine was finally adopted. I commenced its use some time in April, 1885, and my record reads as follows:

"The liquid nourishment recently used was a preparation of the raw expressed juice of beef, rich in albuminoid properties, known as Bovinine, which with milk and eggs furnished a rich, palatable and readily digested diet. This constituted the General's principal article of food during the remainder of his life, he taking it well and with satisfaction up to the 21st of July, when his ability to swallow failed him." J. H. DOUGLAS, M. D. [Physician to General Grant, General McClellan, General Rawlins, and others.]

MY DEAR MR. STUART,   523 Thirteenth Street,
      Baltimore, Md.   Washington, D. C.
I take pleasure in introducing to your favorable notice Bovinine, a nutrient of great value, as I can cheerfully testify from a practical test during my own recent convalescence.

I am sure you will be pleased with this preparation, which presents so many points of improvement over any other meat preparation with which I am acquainted.   I am, yours very truly,
                       D. W. BLISS.
Dr. Bliss was the surgeon in charge of the late President Garfield's case.

               Philadelphia, Pa.
The Bovinine Company,   March 1st, 1887.
GENTLEMEN:—It gives me pleasure to give my testimony to the very great value of Bovinine as a dietetic preparation. I have used it for more than a year in a very aggravated case of nervous dyspepsia, and have found it to answer very much better than any of the many preparations or extracts of meat before used.

I find that it keeps perfectly, even in the warmest weather; is very easily prepared for administration, and it has proved acceptable and beneficial in every case in which I have known it to be given.
         Very respectfully and truly yours,
              R. MURRAY, M. D.,
        Surgeon General (retired) U. S. Army.

     3316 Arch Street, Philadelphia, Pa.
                January, 19th, 1887.
Col. J. H. Baxter, Chief Medical Purveyor, U. S. A., Washington, D. C.

MY DEAR DOCTOR:—I have been asked to say in a letter to yourself what I know about Bovinine. Well, in brief, I will say that my eldest son had capillary bronchitis, and became greatly reduced and exhausted. I put him upon Bovinine, and in the course of two months he regained all his flesh and strength, and has now grown quite robust. I also prescribed it for a niece, who was fragile and delicate in the extreme, in fact neurasthenic, and the result has been that she has now become a strong, healthy woman. Hoping, my dear doctor, that you will pardon this small draft on your valuable time, I am, as ever, your friend,

D. L. MAGRUDER, Surgeon U. S. Army.

Ann Arbor, Aug. 18th, 1886.

GENTLEMEN:—This is to state that I have used Bovinine in a number of cases needing a food at once nutritious and easily assimilated, with very gratifying results.

V. C. VAUGHAN, M. D.

[Professor in the University of Michigan, and discoverer of Tyrotoxicon, &c.]

Louisville, Ky., Oct. 20th, 1887,

GENTLEMEN—I have prescribed Bovinine for several months with satisfactory results. It is the best fluid food in use.

W. H. WATHEN, M. D.

[Dean of Kentucky School of Medicine; Ex-President of State Medical Society, and Consulting Gynecologist to the Louisville City Hospital.]

" Where swift support is called for, for rapid vitalization, there is no better than that form of defibrinated blood advertised as Bovinine. It is fairly palatable, easily assimilated, feels good in the stomach and will keep an indefinite time, the latter quality being absent from all other similar preparations with which I have experimented. If every one had not used Bovinine already and made up his mind as to its value, I should be glad to advise its trial, but it is not needed—suggestion is sufficient." DR. W. F. HUTCHINSON,
in the N. E. Medical Monthly.

MEAT EXTRACTS PREPARED BY HEAT, BEEF TEAS, ETC., HAVE LITTLE OR NO FOOD VALUE, AS WILL BE SEEN FROM THE FOLLOWING OPINIONS.

Many "meat extracts," "essences," etc., are upon the market, a large portion of them containing no food or nutritive properties; for instance, the highly-prized "Beef Tea," popularly supposed to have great food virtues, but actually containing none whatever, being merely a temporary stimulant, or flavor, upon which a patient would soon starve.

In the *British Medical Journal*, Dr. Milner Fothergill says that "a patient dying of exhaustion is generally dying of starvation. We will give him Beef Tea, calf's foot jelly, alcohol, seltzer and milk; that is, a small quantity of sugar and milk, and some fat. But the jelly is the poorest sort of food, and the Beef Tea a mere stimulant. The popular belief that Beef Tea contains 'the very strength of the meat' is a terrible error; *it has no food value whatever.*"

Professor Robert Bartholow, of the Jefferson Medical College, Philadelphia, an undoubted authority in such matters, says: "Nothing has been more conclusively shown than that Beef Tea is not a food. It is nothing more than a stimulant. *All extracts of the nature of Beef Tea carry but a portion of the properties of the meats that they are made of, and in most cases are of no avail as they cannot be retained by the enfeebled stomach.*

Extract from a report on several descriptions of meat preparations, by Dr. A. Stutzer, of Bonn, Director of the Imperial Agricultural and Chemical Laboratory of Rhenish Prussia:

"The beef extract made from Liebig's receipt is not an article of food proper, as containing no more than 7 per cent. of nitrogenous matter; nor did the illustrious professor introduce it as an article of food destined to nourish the human body. *Liebig's idea was not to produce a nutritive article, only a relish.*"

Dr. Stutzer further exposes the often exposed superstition regarding the nourishing powers of Beef Tea. He shows conclusively that one would have to take a half-gallon of Beef Tea before he would get as much nourishment as is contained in a quarter of a pound of steak. He also calculates that a patient would be obliged to consume eighty pints of that deceptive liquid, Beef Tea, prepared from eighty pounds of steak, to obtain the flesh-forming and blood-producing constituents in *one bottle* of Bovinine.

Bovinine containing as it does 20 per cent. of coagulable albumin, is *not* to be classed with preparations of the above character.

[From "*Food and Sanitation.*"— London.]

MEAT EXTRACT REVELATIONS.

In a series of analyses and experiments made in 1891 by Mr. R. H. Chittenden, Professor of Physiological Chemistry at Yale University, the results of which were communicated to the Philadelphia County Medical Association, on May 13th of the same year, Mr. Chittenden gave the following as the percentage composition of Liebig's Extract of Beef, Valentine's Meat Juice, and Bovinine. The figures were:—

	Liebig's Extract of Beef. 2 ozs.	Valentine's Meat Juice. 2 ozs.	Bovinine. 2 ozs.
	Cost 1/2½	Cost 3/-.	Cost 11 d.
Water (at 110 C.)	20·06	60·31	81·69
Solid matter	79·94	39·69	18·91
Inorganic constituents	24·04	11·30	1·02
Phosphoric acid P2 O5	9·13	4·00	0·03
Fat (ether extractives)	0·91	0·78	1·49
Total nitrogen	9·52	2·68	2·43
Soluble albumen coagulable by heat	0·06	0·55	13·98
Total proteid matter available as nutriment	0·06	0·55	13·98
Nutritive value as compared with fresh, lean beef (lean beef = 100)	0·30	2·80	72·40

These analyses were, as regards Liebig's and Valentine's Meat Juice, practically confirmed by results published by Mr. Jesse B. Battershall, Ph.D., F.C.S., chemist, United States Laboratory, New York. Mr. Battershall in FOOD ADULTERATION, page 256, gives the following analyses:—

	Liebig's Extract.	Valentine's Meat Juice.
Water	18·27	50·67
Organic Substance	58·40	29·41
Ash	23·25	11·52
Soluble Albumen	0·05	—
Alcoholic Extract	4·11	—
Phosphoric Acid	7·83	3·76
Potassa	10·18	5·11

Readers of FOOD AND SANITATION do not need to be told that we accept nothing on hearsay or upon testimonials. We believe that for the public safety every food, every drug, and, above all, every patent preparation ought to be subject to analysis, as we are now subjecting every patent food offered to the medical profession and the public. The results, be they favorable or the reverse, ought to be honestly stated, the more so as few medical men have the appliances, the time and the training necessary to make individual in-vestigations. We shall, therefore, in investigating Mr. Chittenden's and Mr. Battershall's analyses, deal with them in a plain, matter-of-fact manner, and entirely upon their merits, however unexpected may be the results and astonishing the conclusions. Divesting thus our minds of cant, we have first to note that Mr. Chittenden gives the soluble albumen as 0·06, whilst Mr. Battershall finds 0·05 per cent. in Liebig's Extract. In Valentine's Meat Juice Mr. Chitten-den finds 0·55, *i. e.*, a little over one-half per cent. On the assump-tion, therefore, that Valentine's Meat Juice is worth as a nutrient 3/-, Bovinine (containing 13·98 per cent. of available nutrient mat-ter) is worth £3 16s. 3d. per bottle of two ounces, although Bovin-ine sells at 2/9 per six ounces (= 11d. for 2 ounces), as against 3/- charged for two ounces of Valentine's Meat Juice. A similar calcu-lation with regard to Liebig's Extract discloses the fact that if Liebig's be worth 1/2½ for 2 ounces, Bovinine is worth, on Mr. Chittenden's analysis, the sum of £14 1s. 6½d. per two ounces.

So much for Mr. Chittenden and Mr. Battershall. Our analyses of both Liebig's and Valentine's preparations show that they are more valuable in nutrients than the analyses above given disclose, and that the results of Mr. Chittenden and Mr. Battershall require correction; the gelatine and albumen in Liebig's being 1·35 per cent., and in Valentine's 0·93 per cent. For the purpose of an ac-curate comparison, we give our own full analysis of the three prepa-rations in a comparative form:—

	Liebig's Extract of Beef. 2 ozs. 1/2½.	Valentine's Meat Juice. 2 ozs. 3/-	Bovinine. 2 ozs 11 d.
Water	16·87	55·24	Water and Alcohol 78·42
Fat (Ether Extract)	3·04	4·80	0·00
Gelatine and Albuminoids	1·35	0·93	13·32
Peptones	8·20	1·55	0·00
Creatine and Meat Extractives, (almost non-nutritious)	47·32	18·27	0·55
Salt	5·08	2·62	1·04
Mineral Matters, Salts of Flesh, Phos-phates, etc.	17·46	8·51	0·57
Non-nitrogenous Extractives	0·68	8·08	6·01
	100·00	100·00	100·00

Calculated comparatively on our own analyses, the albuminoid values work out as follows, taking Valentine's Meat Juice as the basis for the calculations: — Valentine's, containing 0·93 albumin-oids, costs 3s.; Liebig's, with 1·35, costs 1s. 2½ d.; and Bovinine, with 13·32, costs 11d. If, therefore, Valentine's be worth 3s., Liebig's is worth 4s. 4d. instead of 1s. 2½d., and Bovinine is worth 43s. To avoid the possibility of error, we have had duplicate ana-lyses made of Bovinine. The second analysis gave the following results:—

BOVININE.

```
Water and Alcohol..............................................80·70
Fat (Ether Extract)............................................. 1·21
Gelatine and Albumen...........................................12.92
Peptone........................................................ 0·62
Creatine and Meat Extractives.................................. 0·21
Salt........................................................... 0·78
Other Mineral Matters.......................................... 0.08
Non-Nitrogenous Extractives.................................... 3·48
                                                             ———————
                                                             100·00
```

It will be noticed that there is a slight difference in the figures, but this is to be expected in any food preparation, which must vary to some extent. The lesson to be learnt from these analyses, and from the others already published in this series of articles, is, that the system of manufacture of patent food preparations needs revolutionising, and that no medical man can afford to neglect the study of the real value of the meat extracts, etc., he prescribes. At one time the Liebig and Valentine preparations were undoubtedly the best that were placed at the service of the physician. Instead of advancing with the progress of science, the makers have been content to rest where they were when the science of foods was in its infancy. Tradition has invested them with a halo of testimony no doubt honest enough in its time, and with respect to some of the most largely prescribed patent foods, that tradition has been handed from teacher to teacher, from father to son, and from school to school, without anyone being iconoclastic enough to examine into its *bona fides*. It is dangerous for the medical man, above all men, to take anything on trust. Dr. W. M. Grailly Hewitt, in an address to the British Medical Association, emphasised that danger in weighty words, saying:—

"For the last six or seven years I have tested by carefully inquiring into the past history of patients mostly suffering from some uterine or ovarian disease, or some affliction incidental to child-bed, and these conclusions have stood the test of this long-extended inquiry. I have to state the important conclusion that a continuous insufficiency of food, or what may be called a 'chronic starvation,' more or less intense in different cases, was found to have existed universally. Consequently, I have naturally been led to consider chronic starvation as a most important factor in disease."

Experiments have proved that an animal fed upon Liebig's extract of beef alone, will succumb more readily than a like animal entirely deprived of food. The claims advanced, therefore, in favor of preparations of Liebig's Extract of Beef, Meat Juices, and Beef Essences that they represent 30 or 40 times their weight of lean beef is absurd, inasmuch as such preparations are practically devoid of nutritive value, and the waste creatine and extractives (at times half the bulk of the preparation), although of some value as stimulants, have the disadvantage that their use in many diseases is positively dangerous. It is time, therefore, that such preparations were relegated to the kitchen for use as flavoring agents — with which no one would quarrel — and that their makers ceased to represent them to the public as foods. What, for example, are we to think of the following announcement accompanying Brand's Essence of Beef—

"BRAND & CO.'S ESSENCE OF BEEF.

"This essence conists of the juice of the finest beef, extracted by a gentle heat, without the addition of water or any substance whatever, by a process first discovered by ourselves in conjunction with a celebrated physician.

"In cases of extreme exhaustion or urgent danger, a teaspoonful may be administered as often as the patient can take it, in less urgent cases it may be taken as required with a small piece of bread and a little wine."

—in the light of the following analysis:

	Brand's Essence of Beef Cost 1s. 2d.
Water	91·23
Fat (Ether Extract)	0·18
Gelatine and Albuminoids	1·25
Peptones	2·54
Creatines and Meat Extractives almost non-nutritious	3·96
Salt	0.45
Other Mineral Matters	0·39
Non-nitrogenous Extractives	None.
	100.00

or of the assertions *re* Liebig's Extract — that it "makes the most nourishing of beef tea?" Again, if we take the nutrient values of the three preparations, as compared with fresh lean beef —

Reckoning lean beef as equaling	100·00
Liebig's equals only	0·30
Valentine's equals only	2·80
whilst Bovinine equals	72·40

109

FORMULA OF BOVININE.

Defibrinated Bullock's Blood.......................65.00
Desiccated Egg Albumen...........................19.00
Old Bourbon Whiskey..............................10.00
Chemically-pure Glycerine......................... 5.00

___ ___   100.00

ANALYSIS OF BOVININE.

Washington, D. C., Sept. 30, 1887.

The Bovinine Co.,

DEAR SIRS:—A microscopic examination of Bovinine reveals the presence of large quantities of red and white blood corpuscles; also minute fat globules and crystals of leucine and tyrosine. No fibrin or bacteria present. The blood corpuscles are practically unchanged, the red cells being simply decolorized, due to their suspension in a watery medium.

Culture tubes of nutrient jelly, agar-agar, and peptone broth, inoculated with bovinine and kept in an incubator for a week, failed to develop any bacteria.

One or two drops of bovinine placed in a test-tube with 10 c.c. of water, heated, and a drop or two of nitric acid added, reveals the presence of large quantities of albumen.

W. M. GRAY, M. D.,
Microscopist to Army Med. Museum.

J. J. Smith, M.A.F.R.S., Bristol Med. Chir. Journal, Bristol, Eng., Sept. 1, 1887:—

"Bovinine is a dark colored fluid containing little sediment, the amount of coloring matter, estimated by the globinometer, being .20. Its spectrum gives a well marked band at C, with a fainter band at D, resembling that given by an alkaline solution of hæmatin. The fluid is neutral in reaction, and contains 18.2 per cent of coagulable albumen. The nitrogenous extractives soluble in alcohol, and including the inorganic salts (phosphates, sulphates and chlorides of potassium, sodium and calcium), amount to 6 per cent. The nitrogenous extractives insoluble in excess of alcohol amount to 0.45 per cent."

"A fluid containing one-fifth of the coloring matter of blood, with 18 per cent. of extractives and salines, ought to be excellent material from which to manufacture new blood. We are inclined to think that this fluid is likely to become better known and more extensively employed than at present; its dietetic value is undoubtedly great."

OTHER MEDICAL TESTIMONIALS.

(Enough of these might be given to fill another book like this.)

Lexington, Ky., Nov. 2nd, 1887.

GENTLEMEN:—I take pleasure in mentioning that I have given a satisfactory trial of the special properties of Bovinine during my visit to Washington, in attendance upon the International Medical Congress, the patient being an infirm maiden lady of 91 years. Other tonics and foods having entirely failed I presented this, and the result was indeed magical. Subsequent letters from them continue the reports of very satisfactory results, and the friends said thereof, ''Give Bovinine everywhere. Give it by the quart; it is the very best thing in the world.'' L. B. TODD, M. D.

DR. J. P. McGEE, of Memphis, Tenn., in reporting to the Tri-State Medical Association of Tenn., Miss. and Ark., a number of cases of Laparotomy performed by him, while discussing the importance of supporting the system and maintaining the strength of the patient, said: ''The importance of an easily digested concentrated fluid form of food is too evident to need more than simple mention, and in this connection I desire to state that while I am not here to advertise anybody's goods, I have found Bovinine incomparably superior in these cases to any food I have ever used, in that you obtain the largest percentage of nutrition with the smallest residuum.''
Memphis, Tenn., Nov. 7th, 1887.

Cincinnati, Ohio, Dec. 1887.

GENTLEMEN:—Allow me to thank you for calling my attention to Bovinine. It has proved of inestimable value to me in cases of dyspepsia. During its use the pain, tenderness, in fact all the attendant symptoms, have subsided. I believe we have in it a nutrient that is non-irritating, still abundant, and one that will be quickly assimilated, thereby gaining the time so much needed for rest in such cases. It should not be forgotten in wasting diseases and following surgical operations, in which a nutrient possessing the above qualities is of the first importance.

D. W. HARTSHORN, M. D., 124 W. 7th St.

Washington, D. C., Oct. 25th, 1887.

GENTLEMEN:—I have been engaged in the regular practice of medicine for more than 30 years, and in all that time have found nothing that I regard as equal to the Bovinine in all cases indicating its use. I most heartily recommend it to the notice of the profession.

J. STINSON HARRISON, M. D.

Washington, D. C., Oct. 20th, 1887.

GENTLEMEN:—I take great pleasure in adding my testimonial to the remarkable value of Bovinine as a nutriment. I have used it extensively in my practice, and can record many cases where it proved invaluable as an adjuvant in impaired digestion, anæmia with mal-assimilation, phthisis, adynamic fevers, and indeed where there is any loss of tissue I find that I can depend upon Bovinine without fear of disappointment.

M. ESTER HART, M. D.
Attending Phys. and Phar. Homeo. Free Dispensary, Wash., D. C.

Port Huron, Mich., Feb. 6th, 1888.

GENTLEMEN:—I have given your fluid food Bovinine an exhaustive trial, in which its merits were fully tested, and I find it everything you claimed for it. In a practice of forty-seven years I have found nothing to equal it in its power to restore a patient after being very low by wasting diseases.

May you prosper in the production of such an excellent nutrient. I shall never be without it.

WM. CAMPBELL, M. D., 317 Park St.

Toronto, Ont., March 31st, 1887.

GENTLEMEN:—I take great pleasure in eulogizing your Bovinine. I regard it as the very best food for invalids, or those debilitated from any cause, I have ever used; and not only for those sick or convalescing, but for exhausted professional or business men. Bovinine will, better than any nutrient I know of, build up and restore the overtaxed mind and body.

E. T. ADAMS, M. D., 450 Yonge St.

Allegheny, Pa., Jan., 1888.

GENTLEMEN:—I have used your fluid food, Bovinine, and am convinced that it is the best and only reliable food of the kind in the market. I am myself subject to dyspepsia, etc., and have used Bovinine in my own case with excellent results.

A. P. H. SHAFER, M. D.
Late Dist. Phys. of the City Poor Board.

N. B.—This is the first indorsement I have given to any preparation, but your food is so superior to any I have used that I thought it but justifiable in this case. A. P. H. S.